Holding Their Ground

Secure Land Tenure for the Urban Poor in Developing Countries

Edited by
Alain Durand-Lasserve
and
Lauren Royston

Earthscan Publications Ltd
London • Sterling, VA

First published in the UK and USA in 2002
by Earthscan Publications Ltd

ISBN: 1 85383 891 8 paperback
1 85383 890 X hardback

Typesetting by PCS Mapping & DTP, Gateshead
Printed and bound in the UK by CPD Wales, Ebbw Vale
Cover design by Declan Buckley

For a full list of publications please contact:

Earthscan Publications Ltd
120 Pentonville Road, London, N1 9JN, UK
Tel: +44 (0)20 7278 0433
Fax: +44 (0)20 7278 1142
Email: earthinfo@earthscan.co.uk
Web: **www.earthscan.co.uk**

22883 Quicksilver Drive, Sterling, VA 20166-2012, USA

A catalogue record for this book is available from the British Library

Library of Congress Cataloging-in-Publication Data

Holding their ground : secure land tenure for the urban poor in developing countries / edited by Alain Durand-Lasserve and Lauren Royston.
p. cm.
Includes bibliographical references and index.
ISBN 1-85383-891-8 (pbk) — ISBN 1-85383-890-X (cloth)
1. Land tenure–Developing countries. 2. Urban poor–Developing countries. I. Durand-Lasserve, Alain. II. Royston, Lauren, 1967–

HD1131.H65 2002
333.3'1'091724—dc21

2002001292

Earthscan is an editorially independent subsidiary of Kogan Page Ltd and publishes in association with WWF-UK and the International Institute for Environment and Development

This book is printed on elemental chlorine-free paper

Contents

PART 4 CONCLUSIONS

List of Tables, Figures and Boxes

Tables

Figures

Boxes

About the Editors

Alain Durand-Lasserve is a research director at the Centre National de la Recherche Scientifique in France. He is attached to the SEDET Research Centre, University Denis-Diderot in Paris, and specializes in urban land management and policies in developing countries. During the past few years he has been responsible for various research initiatives on tenure regularization policies for irregular settlements in Asia, Latin America, sub-Saharan Africa and Arab countries for the French government and other European bilateral cooperation agencies, multilateral donor agencies and the World Bank. He has organized several international seminars on urban land tenure issues, tenure security and local development, and has published widely on regularization. His recent publications include *Aménagement Foncier Urbain et Gouvernance Locale en Afrique Francophone*, 2000, UNCHS, series Gestion Foncière; *Aménagement Foncier Urbain et Gouvernance Locale en Afrique sub-Saharienne*, 2000, UNCHS, série Gestion Foncière; *Regularisation and Integration of Irregular Settlements: Lessons from Experience*, 1996, Urban Management Programme, Working Paper Series; and *Urban Land Management, Regularisation Policies and Local Development in Africa and the Arab States*, 1995, Urban Management Programme, UNCHS Research and Development Division.

Lauren Royston is a development planner by training and has worked in the urban development field for over ten years. She has worked in the NGO and public sectors and is currently co-principal of Development Works, an urban development organization based in Johannesburg. Her fields of specialization are land and housing and integrated development planning. She regularly undertakes policy and programme assessments, policy advisory and implementation support assignments and action-orientated research projects. Her recent publications include *Urban Land Issues in Contemporary South Africa: Land Tenure Regulation and Infrastructure and Services Provision*, 1998, Development Planning Unit Working Paper No 87, London; 'South Africa: The Struggle for Access to the City in the Witwatersrand Region', in A Azuela, E Duhau and E Ortiz (eds), 1998, *Evictions and the Right to Housing*, International Development Research Centre, Ottawa, forthcoming.

About the Contributors

Cecile Ambert is an associate at Development Works in Johannesburg. She holds an MSc from the University of the Witwatersrand. She specializes in the fields of housing and urban land reform and integrated development planning. Other areas of interest are government institutional and fiscal systems and the impact of HIV/AIDS on development. Recent publications include 'Development, the State and its Leadership in Malawi and Madagascar: The Neoliberal Paradigm Revisited', *Africa Insight*, 1997, No 1; 'Alternatives to Traditional Development Policies: Acknowledging the Links between Security and Development', *African Security Review*, 1997, No 3; and *Participatory Planning in Post-Apartheid South Africa*, Urban Futures Conference, 2000, http://www.wits.ac.za/urbanfutures/papers/ambert.htm.

Banashree Banerjee is an urban planner and most of her work is related to urban poverty, particularly land issues and participatory approaches to poverty reduction. She is an associate staff member of the Institute for Housing and Urban Development Studies (IHS) in Rotterdam and she also works as an independent consultant. Her recent publications related to land issues include 'Security of Tenure: Integrated Approaches in Visakhapatnam', *Shelter*, 1999, Vol 2, Nos 3–4; *Policies, Procedures and Techniques for Regularising Irregular Settlements in Indian Cities*, 1994, Amsterdam Free University, Urban Research Working Papers Nos 34–35; 'Urban Land: Is There Hope for the Poor?', in K Singh and F Steinberg (eds), *Urban India in Crisis*, 1995, Oxford, New Delhi; and 'Three Indian Cases of Upgradable Plots', *Third World Planning Review*, 1994, Vol 16, No 3.

Catherine Cross is based in the Integrated Rural and Regional Development programme at the Human Sciences Research Council in Pretoria. She has published extensively on small farmer development, land reform and land relations in South Africa. She recently designed and carried out studies dealing with small farmer activity in relation to internal migration, carried out three field studies of small farmer activity and capabilities in a rural–urban context in relation to land reform, and undertook qualitative and quantitative field studies in relation to rural credit delivery for small agriculture. She served as research methods consultant to the national review of land reform in South Africa and on the national task team constructing the questionnaire for the recent large-scale review of land reform progress. Most recently, she completed a study of the position of women in relation to land reform delivery.

Colin Davies lives in East London and is the branch land surveyor for the Buffalo City Municipality in the Eastern Cape, Republic of South Africa. He is a registered land surveyor and holds a land management MA (*cum laude*) in land surveying from the University of Natal (Durban). His professional interests are urban informal land tenure, land information systems for informal settlements and managing the linkages between authorities and un-regularized settlements in the context of post-Apartheid South African local government.

Edesio Fernandes is a lecturer at the Development Planning Unit of University College London and co-ordinator of IRGLUS – International Research Group on Law and Urban Space. He is the author of *Law and Urban Change in Brazil*, 1995 (Avebury); and co-editor of *Illegal Cities – Law and Urban Change in Developing Countries*, 1998 (Zed Books).

Clarissa Fourie has a PhD in social anthropology and has worked in the land management/surveying industry for over eight years, as an academic attached to survey departments in South Africa and also as an international consultant. She has worked in both Africa and Southeast Asia for multilateral and bilateral donor agencies, focusing on monitoring and evaluation and the integration of social and technical tools in the land sector. Her recent publications include 'Land and Property Registration at the Cross Roads: A Time for More Relevant Approaches', *Habitat Debate*, 2001, Vol 7, No 3; 'Evaluating Cadastral Systems in Uncertain Situations: A Conceptual Framework Based on Soft Systems Theory', *International Journal of GIS*, forthcoming, 2002, co-authored with Michael Barry; and 'Cadastre and Land Information Systems for Decision-makers in the Developing World', *Geomatica*, 2000, Vol 54, No 3, pp335–342, co-authored with O Nino-Fluck.

Marie Huchzermeyer is a senior lecturer in the School of Architecture and Planning, University of the Witwatersrand in Johannesburg, where she coordinates a postgraduate programme in housing. Her recent publications include 'Housing for the Poor? Negotiated Housing Policy in South Africa', *Habitat International*, 2001, Vol 25, No 3; 'Content and Contradiction: Scholarly Responses to the Capital Subsidy Model for Informal Settlement Intervention in South Africa', *Urban Forum*, 2001, Vol 12, No 1; and 'Informal Settlements: Comparing Production and Intervention throughout 20th Century Brazil and South Africa', *Latin American Perspectives*, forthcoming.

Ellade Imparato is an attorney-of-law with a Masters degree based in the City of São Paulo and working as a consultant on urbanistic law for companies and state agencies on security of tenure, the upgrading of slum housing and land regularization programmes throughout Brazil.

Lysa John coordinates the Urban Policy and Research Unit of Youth for Unity and Voluntary Action. She holds an MA in social work from the Tata Institute of Social Sciences. She is primarily concerned with issues related to urban governance and human rights.

Donald A Krueckeberg is a professor in the Department of Urban Planning and Policy Development at the Bloustein School of Planning and Public Policy and the Center for African Studies, Rutgers University in New Brunswick, New Jersey. His recent publications include 'Private Property in Africa: Creation Stories of Economy, State, and Culture', *Journal of Planning Education and Research*, 1999, Vol 19, No 2; 'The Grapes of Rent: A History of Renting in a Country of Owners', *Housing Policy Debate*, 1999, Vol 10, No 1; 'Property for Everyone and How to Achieve It: The Resident's Property Tax', *Journal of Planning Education and Research*, 1998, Vol 18, No 2; and 'The Difficult Character of Property: To Whom do Things Belong?' *Journal of the American Planning Association*, 1995, Vol 61, No 3.

Kurt G Paulsen is a PhD candidate in the Department of Urban Planning and Policy Development at the Bloustein School of Planning and Public Policy, Rutgers University in New Brunswick, New Jersey. He holds MAs from the University of Wisconsin in development policy and agricultural economics. His research interests are in property and land tenure and local government finance.

Minar Pimple is executive director and co-founder of Youth for Unity and Voluntary Action. He has extensive experience in sustaining communities and creating organizations in urban slum areas as well as building networks and coalitions in the areas of housing, child labour and displacement in India and Asia. He has also planned and conducted international training programmes on human rights, and is author of several papers on the theme of urbanization.

Sonia Rabello de Castro is adjunct professor at the Law School of Rio de Janeiro State University, where she teaches administrative law and public urban law. She also teaches public law on postgraduate courses at many Brazilian universities and research institutes. She is a member of the teaching faculty of Lincoln Institute of Land Policy, participating especially in its Latin American programmes. For the last 25 years she has worked as a lawyer in Brazilian public institutions, first in the Rio de Janeiro Metropolitan Office, then in the Cultural Heritage National Agency, and finally at the Rio de Janeiro Attorney's Office where she was General Attorney. She has published articles and books, in particular on administrative law and urban law, and her latest work is a forthcoming book entitled *Administrative Law, Urban Law and Urban Planning.*

Neelima Risbud is a professor of housing at the School of Planning and Architecture in New Delhi, India. She is an architect, holding an MA in town planning, and she has submitted her doctoral work on 'Slum Improvement and Tenure Regularisation Policies – Indian Context'. Her areas of interest include informal housing and land markets and low income housing projects. Her publications include *Socio-Physical Evolution of Popular Settlements and Government Supports: Case Study Bhopal*, 1987, Research Paper No 3, IHSP; *Illegal Land Supply by Housing Cooperatives and Government Policies – Case Study Jaipur*, Research Study, IHSP; and *Property Market and Settlement Development – Case Study of Rohini Project in Delhi*, 1992, HSMI Delhi.

Nelson Saule is a lawyer, professor of human rights at the Catholic University of São Paulo and a PhD student with an MA in urban rights. He specializes in the fields of housing and urban land reform law. He is president of the Polis Institute of Studies, Formation and Consultancy in Social Policies and a member of the Human Rights Commission of the Order of Brazilian Lawyers (OAB) in São Paulo. His recent publications include *Statute of The City: Guide to Municipalities and Citizens*, 2001, Caica Econômica Federal, Câmara dos Deputados, Presidência da República/Secretaria de Desenvolvimento Urbano, Instituto Pólis, Brasilia; *Right to the City – Legal Ways to the Right to Sustainable Cities*, 1999, Max Limonad, Brazil; and *New Tendency of the Urban Law in Brazil*, 1997, Sergio Antonio Fabris, Brazil.

Acknowledgements

The editors would like to thank the range of institutions and individuals that have given their support in the preparation of this book:

The contributing authors, who submitted their chapters and responded to several rounds of queries and requests for amendments, often under severe pressure of work and mostly at a great distance from the editors and from one another.

All those who contributed papers to the international workshop entitled 'Tenure Security Policies in South Africa, Brazil, India and sub-Saharan Africa: A Comparative Analysis', held in July 1999 in Johannesburg, South Africa. This book contains an edited version of selected papers presented and discussed during the workshop.

The United Nations Human Settlements Programme (UN–Habitat), the Lincoln Institute of Land Policy (LILP) and the French Institute of South Africa (IFAS), whose generous support facilitated the dissemination of our research findings.

IFAS support also facilitated our early collaboration and initial report, and was responsible for the organization of the Johannesburg workshop.

Support from LILP made it possible for our Brazilian colleagues to attend the Johannesburg workshop and was of enormous help to us in our editorial endeavours.

We are extremely grateful for the additional support from the Global Campaign for Secure Tenure, spearheaded by UN–Habitat, which made it possible for us to publicize this book widely and to inform audiences in India, Brazil and South Africa about it.

Kate Shand and Louise Oppenheim calmly and competently assisted us with the editorial work.

The International Research Group on Law and Urban Space (IRGLUS) and the Centre for Applied Legal Studies (CALS) were kind enough to agree to join forces and hold their workshop on redefining property rights in the same week. We are especially grateful to Edesio Fernandes and Theunis Roux for this collaboration.

The Centre National de la Recherche Scientifique, France, and Development Works, Johannesburg, supported our individual involvement in the project, allowing us to devote considerable time and energy to working on it.

List of Acronyms and Abbreviations

ACHR	The Asian Coalition for Housing Rights
ANC	African National Congress
APSHB	Andhra Pradesh State Housing Board
APSHCL	Andhra Pradesh State Housing Corporation Ltd
BDA	Bangalore Development Authority
BIT	Bombay Improvement Trust
BMC	Bombay Municipal Corporation
CBD	central business district
CBO	community-based organization
CHIS	Chinagadili Habitat Improvement Scheme
CMA	Cape Metropolitan Area
CMDA	Chennai Metropolitan Development Authority
COMUL	Comissão de Urbanizacão e Legalizacão (Commission for Upgrading and Legalization)
DDA	Delhi Development Authority
DFA	Development Facilitation Act
DFID	Department for International Development (UK)
DLA	Department of Land Affairs
DMC	Deputy Municipal Commissioner
DTP	Director of Town and Country Planning
EIUS	environmental improvement of urban slums
ESTA	Extension of Security of Tenure Act
FAPESP	Fundação de Amparo à Pesquisa do Estado de São Paulo
FGTS	Fundo de Garantia de Tempo Serviçio (Guaranteed Fund for Time Employed)
FIG	International Federation of Surveyors
FSI	Floor Space Index
GEAP	Grupo Executivo para Assentaments Populares (Executive Group for Popular Settlements)
GEAR	Growth, Employment and Redistribution strategy
GOAP	Government of Andhra Pradesh
GPS	global positioning system
GUD	guided urban development
HIC	Habitat International Coalition
HUDCO	Housing and Urban Development Corporation
IDT	Independent Development Trust
ISLP	Integrated Serviced Land Project

JRS	Jogeshwari Rahiwasi Sangh
LIS	land information system
MCD	Municipal Corporation of Delhi
MCGM	Municipal Corporation of Greater Mumbai
MCV	Municipal Corporation of Visakhapatnam
MHADA	Maharashtra Housing and Area Development Authority
MMRDA	Mumbai Metropolitan Region Development Authority
MUDP	Madras Urban Development Project
MUTP	Mumbai Urban Transport Project
MVLA	Maharashtra Vacant Lands Act
NDPC	National Development and Planning Commission
NGO	non-governmental organization
NHB	National Housing Bank
NHC	neighbourhood housing committees
NHF	National Housing Forum
NIT	Nagpur Improvement Trust
NSDP	National Slum Development Programme
PHIS	Programma Habitaçion de Interesse Social (Social Interest Housing Programme)
RDP	Reconstruction and Development Programme
SANCO	South African National Civic Organization
SBPE	Sistema Brasileiro de Poupança e Empréstimos (Brazilian Savings and Loan System)
SJSRY	Swarna jayanti shahari rojgar yojina (Golden Jubilee Urban Employment Scheme)
SPARC	Society for Promotion of Area Centres
SRD	slum redevelopment scheme
SRS	slum rehabilitation scheme
SUP	slum upgrading programme
TA	Tribal Authority
TCPO	Town and Country Planning Organization
TLC	transitional local council
TNUDP	Tamil Nadu Urban Development Project
UBS	Urban Basic Services programme
UBSP	Urban Basic Services for the Poor
UCD	urban community development
UCDD	Urban Community Development Department
ULCRA	Urban Land (Ceiling and Regulation) Act
ULTRA	Upgrading of Land Tenure Rights Act
UNCHS	United Nations Centre for Human Settlements (Habitat)
UNICEF	United Nations Children's Fund
USAID	US Agency for International Development
VLT	vacant land tenancy
VMR	Visakhapatnam Metropolitan Region
VPT	Visakhapatnam Port Trust
VSIP	Visakhapatnam Slum Improvement Project

VUDA	Visakhapatnam Urban Development Authority
WHO	World Health Organization
YUVA	Youth for Unity and Voluntary Action
ZEIS	Zonas Especiales de Interesse Social (Zones of Special Social Interest)

Chapter 1

International Trends and Country Contexts – From Tenure Regularization to Tenure Security

Alain Durand-Lasserve and Lauren Royston

BACKGROUND

Urban land issues and tenure security are emerging as one of the major challenges for decision-makers, planners, professionals and researchers involved in urban management and in the implementation of new land and housing policies for the urban poor. This book presents some of the main findings and conclusions of a comparative research programme on land tenure issues and access to land for the urban poor in developing cities. It was initiated during the preparation of the Second United Nations Conference on Human Settlements (Istanbul, 3–14 June 1996) and continued until the present by a large group of experts and researchers in Asia, Latin America and Africa. The programme has been supported by various multilateral and bilateral aid and cooperation agencies, especially through the Urban Management Programme.

In its initial stage, the research programme focused on the concepts, procedures and techniques involved in the regularization of irregular settlements, with particular attention being given to tenure regularization policies. A series of comparative analyses were carried out in Asia, Latin America and sub-Saharan African countries, and were presented at two international seminars held in 1993 and 1995 (Urban Management Programme, 1993 and 1995b). These initiatives underlined the key role of security of tenure in all policies aiming to improve access of the urban poor to housing and urban services.

Several parallel research initiatives in the 1990s contributed to the growing concern for security of tenure issues. By the late 1990s, security of tenure was becoming one of the major issues facing officials in charge of planning and housing programmes, especially irregular settlement upgrading programmes. It topped the agenda following the Habitat II preparatory conference on 'Access to Land and Security of Tenure as a Condition for Sustainable Shelter and Urban Development', held in New Delhi in 1996 (UNCHS, 1996c). The conference declared an initial commitment from governments at central and local levels, the private sector and the community sector as well as from a broad range of organizations representative of civil society. The New Delhi Declaration (UNCHS, 1996a) recommended that 'security of tenure should be given to inhabitants of irregular settlements, slums, shack dwellers, and all precarious living environments'. Following this, the Habitat II conference debates and conclusions underlined the importance of tenure issues in the implementation of the United Nations Centre for Human Settlements (Habitat) (UNCHS) Habitat Agenda (UNCHS, 1996b), and specifically security of tenure. Governments committed themselves to the objective of 'Providing legal security of tenure and equal access to land to all people, including women and those living in poverty' (UNCHS, 1996b, paragraph 40).

> *Access to land and security of tenure are strategic prerequisites for the provision of adequate shelter for all and for the development of sustainable human settlements affecting both urban and rural areas. It is also one way of breaking the vicious circle of poverty. Every government must show a commitment to promoting the provision of an adequate supply of land in the context of sustainable land use policies. While recognizing the existence of different national laws and/or systems of tenure, governments at appropriate levels, including local authorities, should nevertheless strive to remove all possible obstacles that may hamper equitable access to land and ensure that equal rights of women and men related to land and property are protected under the law. The failure to adopt, at all levels, appropriate rural and urban land policies and land management practices remains a primary cause of inequity and poverty* (ibid, paragraph 75).

In 1999, UNCHS decided to focus its energy on a limited number of key issues. As a result, two global campaigns were launched, one of which is security of tenure (the other being governance) (UNCHS, 1999a).

This was the global context for the organization of an international workshop entitled 'Tenure Security Policies in South Africa, Brazil, India and Sub-Saharan Africa: A Comparative Analysis' in July 1999.

Aims and content of the book

This book contains edited versions of selected papers on security of tenure policies in South Africa, Brazil and India presented and discussed during the workshop. In this general introduction, in the country introductions and in the general conclusion, the contents of the debates around these papers are described, and major issues relating to the implementation and enforcement of security of tenure policies are identified. The main components, characteristics and achievements of these security policies in South Africa, Brazil and India are analysed and a link established between these policies and the ongoing theoretical debate on security of tenure issues at an international level.

The book looks at how pragmatic solutions can be found and implemented to respond to the demands and needs of the majority of urban households living in informal settlements, and analyses how urban stakeholders, under particular social, legal and economic constraints, are devising and implementing innovative and flexible responses.

It is hoped that this publication will fill a gap in comparative research on tenure upgrading policies and document the ongoing debate on tenure issues for the urban poor in developing cities. It also aims, first, to consolidate relationships between researchers, professionals and decision-makers involved in defining and implementing tenure upgrading policies and programmes in South Africa, Brazil, India and sub-Saharan African countries; and, second, to contribute to the UNCHS Global Campaign for Secure Tenure (2000–2001). It primarily addresses researchers and academics, students, professionals involved in urban tenure upgrading programmes, officials in charge of urban land management and development aid agencies.

IRREGULAR SETTLEMENTS AND SECURITY OF TENURE

The spatial growth of irregular settlements in cities in developing countries reflects increasing disparities in the distribution of wealth and resources. All empirical studies carried out during the last ten years have emphasized the magnitude of the problem. In most developing cities in Asia, Latin America, sub-Saharan Africa and the Arab States, between 25 and 70 per cent of the urban population is living in irregular settlements, including squatter settlements, unauthorized land developments, and rooms and flats in dilapidated buildings in city centre areas (Durand-Lasserve, 1996 and 1998). Although the situation varies widely from one country to another, the majority of households living in irregular settlements have no formal security of tenure and poor access – if any – to basic urban services.

With very few exceptions, trends indicate a steady deterioration in the tenure status and housing conditions of the urban poor (UNCHS, 2001, pp3–22). Public and private formal land and housing delivery systems simply cannot respond to the needs of the urban poor (UNCHS, 1996a; *Habitat Debate*, 1997). Despite the multiplicity of poverty alleviation initiatives and safety-net programmes, the total number of people living in informal settlements is still tending to increase at a faster rate than the urban population. As a result, informal settlements frequently account for more than 50 per cent of the spatial growth of developing cities.

In most developing cities, the expansion of informal settlements over the last two decades took place in a context of accelerated globalization and structural adjustment policies, combining deregulation measures, privatization of urban services, massive state disengagement in the urban and housing sector, and attempts to integrate informal markets – including the land and housing markets – into the sphere of the formal market economy (World Bank, 1991,

1993; Harris, 1992). These policy measures, along with the lack of, or inefficiency of, corrective measures or safety-net programmes have tended to further increase inequalities in wealth and resource distribution at all levels (Pugh, 1995, pp34–92; Durand-Lasserve, 1994).

As a result, the urban poor and large segments of the low and low to medium income groups have no choice but to rely on informal land and housing markets for access to land and shelter (Skinner et al, 1987; Hardoy and Satterthwaite, 1989). This situation has led to the rapid spatial expansion of irregular settlements. Informal land and housing delivery systems remain the only realistic alternative for meeting the needs of low-income households (Mathey, 1992).

Public policies aiming to integrate irregular settlements within the city have been undermined by the stagnation or decrease in the revenue of many urban households, and growing inequalities. This situation has resulted in accelerated social exclusion and spatial segregation. The rising cost of urban land in most developing cities over the last two decades has further aggravated the situation (Jones and Ward, 1994). As stressed by most observers, shelter conditions for the world's urban poor are not improving. Market mechanisms, as well as the lack of protection and appropriate regulatory measures, are tending to consolidate this trend.

Tenure insecurity, threat of forced eviction and poor access to basic urban services, all contribute to further undermine the economic situation of the poor (Fondation pour le Progrès de l'Homme, 1992; Audefroy, 1994).

Residential tenure status: three main forms of irregular settlements

The need for security of tenure concerns all those living in informal, irregular or illegal settlements. These three terms refer to the same phenomenon. In this introduction the term 'irregular settlement' is used. A wide range of residential tenure systems and practices exist. Some are formal or legal, others are not. Many are 'in between', with some of the attributes of legality, but not all (Payne, 1997).

The main type of irregular settlement is unauthorized land development. Various other terms are used, depending on the country or author, such as: illegal commercial land subdivision, informal subdivisions, informal land developments, *loteamentos* (Brazil), and *colonias* (Mexico). Unauthorized land developments are a widespread phenomenon on the fringes of most developing cities. Most often, such settlements have developed on private agricultural land, frequently outside the municipal boundaries (UNESCAP, 1990). In most sub-Saharan African cities, customary owners are the main providers of land for housing, even if their right to the land is not formally recognized by the state (UNCHS, 1999c).

Occupants have purchased the land – or sometimes rented it – from an informal developer, an individual or communal landowner, either directly or through an intermediary. The sale of the plot is generally legal, although not always registered, but the land may not be considered suitable for urban development or housing, or the development may not comply with planning laws and regulations, or with norms and standards regarding infrastructure and services (Durand-Lasserve and Clerc, 1996). Householders who consider themselves to be owners usually occupy such settlements. The large-scale existence of unauthorized land development for rent seems to be more frequent in cities that are facing the pressure of demand for land for housing from new migrants with very low income levels.

Squatter settlements are found on the urban fringes or in centrally located areas, mostly on public land but also, less frequently, on private land, especially when disputed. These can be the result of an organized 'invasion', or a gradual occupation. Contrary to common belief, access to squatter settlements is rarely free. An entry fee must generally be paid to an intermediary, or to the person or group who exerts control over the settlement, and sometimes also rent. Every country, even every city, has its own term for squatter settlements.

Informal rental housing covers a wide range of situations and levels of precariousness. Rental is the most common form of tenure in formal as well as in informal settlements (Gilbert, 1987; Gilbert and Varley, 1990). Tenants and subtenants form a heterogeneous group. They can be found in unauthorized land developments, in squatter settlements or in dilapidated buildings in city centres.

The social structure of irregular settlements is far from homogeneous within a single city or even within one settlement. Irregular settlements are not always occupied exclusively by the urban poor (Fernandes, 1996). Middle income households settle in these areas when the formal housing market cannot meet their demands and in such cases a certain 'right to irregularity' may be recognized; the situation is periodically set to rights through mass regularization using legal measures. However, in general terms, the map of urban poverty can be superimposed on that of irregular settlements with a fair degree of accuracy.

Considering population dynamics at a global level, the total number of squatters is tending to decrease in most developing cities, whereas unauthorized settlements are on the increase. This closely follows the trend that has been observed for almost two decades; there is no longer free access to land. The main reason for this is the ever-increasing commodification of land delivery systems for the poor of the cities and the fact that there is less and less public land available for occupation by squatters (Baross and van der Linden, 1990; Jones and Ward, 1994).

If we look at the full range of irregular or informal tenure options for the urban poor, there is a great deal of interaction between formal and informal land delivery systems, and between subsystems within each of these systems. It is important to note that when an informal land delivery subsystem becomes saturated, or is no longer able to operate to produce internal or external results, then the demand for land shifts to another subsystem which is supposed to offer comparable advantages at an equivalent cost (Urban Management Programme, 1995b).

The need for secure tenure

Populations living in irregular urban settlements are all confronted with the same set of interrelated problems: they have no access – or limited access only – to basic services, and they have no security of tenure. Their situation is precarious as they usually belong to the poorest segment of the urban population (Cobbett, 1999). However, it must be stressed that informality does not necessarily mean insecurity of tenure.

Some forms of residential tenure arrangement can guarantee a reasonably good level of security. This is the case, for example, in sub-Saharan African countries, in communal or customary land delivery systems, even when these are not formally recognized by the state (Urban Management Programme, 1995b). Recognition by the community itself and by the neighbourhood is often considered more important than recognition by public authorities for ensuring secure tenure. However, this arrangement can deteriorate under some circumstances, for instance:

- when the customary system is in crisis, with leadership conflicts within the group of customary owners, especially between those who allocate the land and others members of the group (Rochegude, 1998);
- when multiple allocations of the same plot generate a series of conflicts within the community (this may be the result of illicit land sales by unauthorized persons, a common phenomenon in the absence of any land information and record system);
- when a major conflict arises between customary owners and public authorities about the ownership and use of the land, or about the legitimacy of the customary claim. In such cases, alliances often develop between customary owners and the community against the public authorities (Rakodi, 1994; UNCHS, 1999c).

Unauthorized land development on private land offers various levels of protection, depending on the public authorities' perception of the degree of illegality of the settlement. Even if the area is not suitable for residential development, occupants can generally produce a deed of sale or a property title for the land they occupy. It is worth noting that, in such settlements, middle and upper-middle income groups are well protected against forced eviction, because of their political influence and their cultural and economic capacity to regularize their situation (International Federation of Surveyors and UNCHS, 1997).

Squatter settlements are more exposed to forced evictions, especially those located on private land in prime urban areas that are therefore subject to high market pressures, and those which occupy hazardous or dangerous sites. The poorest communities are especially vulnerable to external pressures. Frequently there is a lack of any internal cohesion in these settlements, making it difficult for the populations to group together to defend themselves (UNCHS, 1999a).

Whatever the type of irregular settlement (unauthorized land development on customary or private land or squatter settlements on public or private land), four main factors contribute to the protection of households from eviction:

1 the length of occupation (older settlements enjoy a much better level of legitimacy, and thus of protection, than new settlements);
2 the size of the settlement (small settlements are more vulnerable than those with a large population);
3 the level and cohesion of community organization; and
4 the support that concerned communities can get from third sector organizations, such as non-governmental organizations (NGOs).

The situation is different for tenants and subtenants, whether in unauthorized settlements, squatter settlements, dilapidated buildings in city centres or formal settlements (Mitlin, 1997). These are the most vulnerable groups, especially when they are exposed simultaneously to different levels of informality (eg when the owner is herself/himself in an irregular situation) (Laubé, 1994). Except in large and homogeneous rental settlements, such as the shack farming settlements on South African urban fringes, tenants are scattered throughout irregular settlements with a wide range of informal rental arrangements, and they are often unable to organize as a pressure group to protect themselves. They are exposed to the arbitrary decisions of their land or shelter owner, and generally have no recourse to legal advice. Being the poorest among the urban poor, they are unable to meet the costs incurred by any improvement of their living environment (UNCHS, 1993b). Unlike most irregular settlement occupants, they cannot apply for compensation in the case of forced removal and they are generally not eligible for resettlement.

Studies on the socio-economic situation of households living in irregular settlements indicate a strong correlation between urban poverty, tenure status, access to services and citizenship (Vanderschueren et al, 1996; Durand-Lasserve and Clerc, 1996; UNCHS, 1999b). Tenure status is one of the key elements in the poverty cycle. Lack of security of tenure hinders most attempts to improve shelter conditions for the urban poor, undermines long-term planning and distorts prices for land and services (Wegelin and Borgman, 1995). It has a direct impact on access to basic urban services and on investment at settlement level, and reinforces poverty and social exclusion (UNDP, 1991). As underlined by UNCHS (1999a), it impacts most negatively on women and children. In this context, protection against eviction is essential as not only do evictions hit primarily the poorest communities, but they never achieve their officially declared objectives. Rather than solving the problem, they tend to compound and worsen it.

Security of tenure is a key element for the integration of the urban poor into the city

Land tenure refers to the rights of individuals or groups in relation to land. The exact nature and content of these rights, the extent to which people have confidence that they will be honoured, and their various degrees of recognition by the public authorities and communities concerned will have a direct impact on how land will be used (Fourie, 1999). As Fischer (1995) noted:

> *tenure often involves a complex set of rules, frequently referred to as a 'bundle of rights'. A given resource may have multiple users, each of whom has particular rights to the resource. Some users may have access to the entire 'bundle of rights' with full use and transfer rights. Other users may be limited in their use of the resources (ie nature of the use … length of use, etc).*

It is important to bear these words in mind since they underline both the diversity of rights to land and the existence of a wide range of options, from full ownership to less exclusive forms of possession and use. There is a possible coexistence in one place of forms of tenure that give access to different rights and a continuum between these different forms of tenure. This highlights the fact that ownership is only one form of tenure among many others (Payne 1997, 1999b; Fourie, 1999).

What do we mean by 'security of tenure'?

The definition of security of tenure proposed by UNCHS is as follows:

> *Security of tenure describes an agreement between an individual or group to land and residential property which is governed and regulated by a legal and administrative framework. This legal framework is taken to include both customary and statutory systems. The security derives from the fact that the right of access to and use of the land and property is underwritten by a known set of rules, and that this right is justifiable. The tenure can be affected in a variety of ways, depending on constitutional and legal framework, social norms, cultural values and, to some extent, individual preference. In summary, a person or household can be said to have secure tenure when they are protected from involuntary removal from their land or residence, except in exceptional circumstances, and then only by means of a known and agreed legal procedure, which must itself be objective, equally applicable, contestable and independent. Such exceptional circumstances might include situations where physical safety of life and property is threatened, or where the persons to be evicted have themselves taken occupation of the property by force or intimidation* (UNCHS, 1999b).

Thus security of tenure does not necessarily require the provision of leasehold or freehold titles. It can be achieved through other procedures and arrangements.

Protection against forced evictions is a prerequisite for the integration of irregular settlements into the city. For households living in irregular settlements, security of tenure offers a response to their immediate problem of forced removal or eviction. It means they cannot be evicted by an administrative or court decision simply because they are not the owner of the land or house they occupy, or because they have not entered into a formal agreement with the owner, or do not comply with planning and building laws and regulations. It also means recognizing and legitimizing the existing forms of tenure that prevail among poor communities, and creating space for the poorest populations to improve their quality of life. Security of tenure can be considered the main component of the right to housing, and an essential prerequisite for access to citizenship, as was emphasized by the Habitat Global Campaign for Secure Tenure:

> *security of tenure is a fundamental requirement for the progressive integration of the urban poor in the city, and one of the basic components of the right to housing… It guarantees legal protection against forced eviction… The granting of secure tenure is one of the most important catalysts in stabilizing communities, improving shelter conditions, encouraging investment in home based activities which play a major role in poverty alleviation, reducing social exclusion, and improving access to urban services* (UNCHS, 1999b).

However, as most studies have stressed, security of tenure is not in itself sufficient to break the poverty cycle. It forms only a part of a more comprehensive and integrated approach to informal settlement upgrading, as all the case studies presented in this book confirm.

FROM TENURE REGULARIZATION TO SECURITY OF TENURE: THE ONGOING DEBATE ON STRATEGIES REGARDING TENURE ISSUES FOR THE URBAN POOR

The current preoccupation with security of tenure issues of institutions responsible for urban land management and housing development programmes is, to a large extent, the result of lessons learnt from the experience of the last decade. Figures and trends regarding irregular settlements, evaluation of the impact of the tenure status on the poverty cycle and the limited achievements of conventional responses, all suggest the need for a drastic shift in our approach to land and housing issues (UNCHS, 1996a).

Can we identify the emergence of a consensus on tenure issues in the context of developing cities? In her introduction to the Indian case studies (Chapter 2), Banerjee writes that:

> *diverse, and even opposing, factors from within and outside the government machinery have converged and led to consensus on the tenure issue. These are pressures from the civil society and courts, structural adjustment, lessons from experience, international thinking and political agendas.*

Although this assumption cannot be generalized to other countries, it does reflect the trends and changes regarding tenure issues that are currently occurring at a global level, including Brazil and, to a lesser extent, South Africa. Governments are increasingly aware of the limitations of conventional responses based exclusively on tenure regularization programmes.

Conventional responses to irregularity

Conventional responses regarding access to land and housing for the urban poor have been well documented. They are based mainly on the regularization of irregular settlements, emphasizing tenure legalization and the provision of individual freehold (Dowall and Giles, 1991). Approaches combining tenure legalization and titling programmes with programmes to provide serviced land, upgrading and improvements at settlement level have had limited success (Urban Management Programme, 1993 and 1995b). It is now more and more frequently acknowledged that such programmes must be drastically redefined and reassessed (Ward, 1998). This point deserves clarification.

The cornerstone of regularization policies as implemented in some developing countries – such as Mexico during the 1990s (Azuela, 1995; Varley, 1999) – was the massive provision of individual freehold titles or other forms of real rights (a right that can be transferred, inherited and mortgaged) such as long-term leases. Such responses require a series of complex procedures to identify the holders of rights and their beneficiaries, to resolve disputes, to delineate plots by surveying, to pay out compensation if required and to provide land registration and titling (UNCHS, 1991). Although this gives beneficiaries sound security of tenure, it is an expensive and time-consuming process, especially in contexts where the processing capacity of the administrations involved is limited, where land-related information is out of date or insufficient, and where centralized land registration procedures are complicated. Frequent incidents of corruption in administrations in charge of land management and allocation and the low level of literacy among the populations concerned further aggravate the situation.

Conventional responses include the following:

- Tolerance by the public authorities of the existence of a dual, formal/informal, land delivery system (this is the case in most sub-Saharan African countries), but the absence of a clear strategy regarding irregular settlements. They may combine repression (forced eviction, harassment and various forms of pressure), tolerance (laissez-faire policies) and selective tenure regularization, according to the political context. It must be noted that there is always, in principle, a legal procedure that allows individual tenure regularization (Serageldin, 1990).
- Attempts to adapt land law to the situation and needs of developing cities (McAuslan, 1998).
- Formal recognition and legitimization of the existence of informal land delivery systems only when they are considered as being controlled by customary owners, in specific areas and under specific conditions. Most decisions by customary owners must be approved or authenticated by public authorities (as in, for example, Ghana, Uganda, Botswana and Cameroon) (Mabogunje, 1992; Mosha, 1993).
- Reduction of planning and construction norms and standards that act as constraints (Dowall, 1991).
- Integration of informal land and housing delivery systems into the sphere of formal activities through large-scale registration and tenure upgrading

and legalization programmes (as in, for example, Mexico and Egypt) (Varley, 1994; Azuela, 1995; Arandel and El Batran, 1997).
- The setting up of a parallel, alternative system, supposedly simpler and cheaper than the existing formal registration system. This may be based on simplified recording procedures. The entities in charge provide titles that it is possible to mortgage. However, the mortgage value of such titles is less than that of freehold titles (Zimmermann, 1998; Zevenbergen, 1998). A response such as this, which was promoted in Peru by the Instituto Libertad y Democracia, tends to consolidate the dual approach to tenure.
- Tentative, top-down land policy and institution reforms (Farvacque and Mc Auslan, 1992; Von Einsiedel, 1995).

Recent shifts have focused on the following practices:

- Setting up a simplified registration system where tenure can be incrementally upgraded to real rights in accordance with the needs and resources of individual households and the processing capacity of the administrations in charge (as with the experience in Namibia) (Christiensen and Hoejgaard, 1995; United Nations Economic Commission for Africa, 1998; Christiensen et al, 1999; Fourie, 1999). A system such as this must be compatible with formal registration procedures.
- Devising and adopting innovative tenure formulae that emphasize collective trust or cooperative ownership. In the context of most cities this is an appropriate though temporary solution that has difficulty in resisting market pressures. This practice has limited capacity for innovation.
- Emphasizing partnership between formal and informal actors (Payne, 1999a).
- Emphasizing security of tenure (protection against evictions) whenever possible, through long-term lease and/or other measures which, first, give priority to the consolidation of occupancy rights (the right not to be evicted, to have access to basic urban services, etc) rather than to the provision of property or freehold titles, and, second, give priority to collective rather than individual interests. In different cities these basic responses can be combined in different ways (UNCHS, 1996b; Payne, 1999a; Durand-Lasserve, 2000).

Accompanying measures are usually adopted in order to facilitate the implementation of these responses. Such measures may include the setting up of appropriate Land Information Systems (LIS), including simplified multipurpose cadastres (UNCHS, 1993a), or improving and updating existing ones, and the promotion of conflict resolution procedures, land adjudication procedures and survey techniques adapted to the tenure situations of irregular settlements (Durand-Lasserve, 1993; Dale, 1997; International Federation of Surveyors and UNCHS, 1997; Williamson, 1998). Here again, recent shifts indicate a new approach to tenure issues, with emphasis mainly on the following (UNCHS, 1999c; Fourie, 1999):

- decentralization of land management responsibilities to local/municipal levels with municipalities receiving sufficient resources (both human and financial) to carry out land registration and land allocation and use (Rakodi, 1999);
- attempts at integrating legal pluralism approaches into tenure policies (Tribillon, 1993; Benton, 1994);
- reliance on community-based and grass-roots organizations at settlement and city levels (Abbott, 1996; Imparato and Ruster, 2001);
- provision of basic services as a form of settlement recognition and as a tool for alleviating poverty;
- improved access to credit for the urban poor, through conventional and micro-finance systems (Aurejac and Cabannes, 1995).

Despite recent shifts towards more flexible tenure regularization procedures, emphasis is still placed on access to individual ownership based on the allocation of individual property titles. The underlying model remains based on the formal/informal dichotomy and underestimates – or does not even take into account – the diversity and legitimacy of other tenure arrangements and the existing continuum between tenure systems. As noted by Payne (1999b):

> *The range and complexity of tenure systems ... demonstrates that it is simplistic to think of tenure in black and white terms of legal or illegal, since there is generally a continuum of tenure categories within most land and housing markets. In many countries, there may even be more than one legally acceptable system operating, so that migrants moving from customary areas to urban centres may be considered to be behaving illegally, simply because they are operating in accordance with systems which are not locally applicable. The coexistence of these different tenure systems and submarkets within most cities creates a complex series of relationships in which policy related to any one has major, and often unintended, repercussions on the others.*

Culturally and ideologically oriented development models

The emphasis that most countries still place on options favouring private land and housing ownership, to the detriment of other options, is due largely to conventional responses to the expansion of informal settlements which always reflect culturally and ideologically oriented development models (Rolnick, 1996; Feder and Nisho, 1998). Diagnoses of the current situation regarding access to land and housing, perception of needs and rights and responses are mainly guided by forms of technical rationality and financial logic that have been designed by international finance institutions and aid agencies (Torstensson, 1994). As part of the measures aimed at improving the management of urban institutions, improving local resource mobilization and establishing enabling regulatory frameworks, the World Bank's strategy regarding urban development in the 1990s (World Bank, 1991; Cohen and Sheema, 1992) advocates the improvement of land information and registration systems and the introduction of regulatory reforms in order to improve the functioning of urban land markets. The strategic role of market-oriented urban land and housing policies was repeatedly emphasized by the World Bank during the 1990s (World Bank, 1991 and 1993). In this context, priority is given to tenure regularization of irregular settlements and to the upgrading of land tenure systems. The long-term objective is to promote private ownership through the allocation of individual freehold/property titles. This may have a negative impact on the urban poor. As Payne (1999b) notes, the World Bank is:

> *surprisingly reticent regarding the impact of its tenure proposals on the rental sector, particularly private informal rental housing, which accommodates a large proportion of the urban population and almost all of the poorest households. There is therefore a real danger that a policy approach, which emphasizes the benefits of owner occupation, and provides various incentives for it, may result in the creation of a large underclass that is denied access to any form of affordable or acceptable housing. This fails to take into adequate account the variety of legal and socially accepted traditions in land tenure systems and distorts land markets in favour of one system at the expense of all others. This is hardly consistent with the objective of improving the equity of urban land and housing markets… The important point is that policies, which emphasize and encourage freehold, may unintentionally or inadvertently discriminate against other forms of tenure that may be more appropriate for large sections of the population. For example, it is common for many low-income households to prefer the mobility offered by rental tenure systems, provided they enjoy adequate security and legal rights. Such protection may be easier to achieve in land markets which encourage a variety of tenure options, rather than one at the expense of others.*

One of the basic hypotheses behind urban land policies in general, and tenure reforms in particular, is still that home ownership and the provision of property titles is the only sustainable solution for providing security of tenure to the urban poor while facilitating the integration of informal land markets within the framework of the formal economy. The convergence of diagnoses and responses has as its starting point a similarly converging analysis of the role of the city in economic development (World Bank, 1991), and the certainty that an increase in urban productivity would result from the unfettered development of the market economy through privatization, deregulation, decentralization and improvements in the financial systems. So-called corrective measures and safety-nets are supposed to lessen the social and environmental effects of these policies.

Such a convergence is illustrated, at a global level, by the adoption of a standardized vocabulary for and reference to the same notions and concepts (such as productivity, efficiency, deregulation and privatization). This vocabulary is by no means neutral. Relations between urban stakeholders – including tenure relations – are seen mainly as being organized around the supply and demand relationship, a relationship that the World Bank has gradually formalized since the beginning of the 1980s. As a political scientist stated recently (Hibou, 1998), one of the main principles underlying this discourse is the 'will to curb politics, while reinforcing the choice of liberal economy standards and the search for simplicity'. In order to get round the problem of politics, international institutions 'have called on political economy theories which tend to depoliticize perceptions and interpretations'. Political actors are analysed as economic actors. This predominant discourse 'reflects a political and moral position which uses watered down scientific theses to legitimize itself'.

Aid agencies are increasingly questioning urban land policies based exclusively on access to individual ownership. The World Bank itself is now tending to focus less on freehold and instead pay more attention to security of tenure issues:

> *The issues of involuntary resettlement have become more intense in many cities as population pressure and land market rigidities are provoking further encroachment on environmentally vulnerable lands and rights of way. Fair and sustainable resettlement programmes need to be designed with due regard to citywide land market policies and conditions, not as neighbourhood enclave activities, since displaced households will often return to squatter status elsewhere. The cost and availability of alternative housing sites, access to employment, and security of tenure are key considerations for the welfare of resettled households and therefore for the design of workable resettlement programmes* (World Bank, 1999).

Providing security of tenure

Responses to tenure insecurity vary according to local contexts, to the types and diversity of irregular settlements, to governments' political orientations, and to pressures from civil society in general and from the communities concerned in particular.

There are two main approaches, which differ but are not contradictory. The first emphasizes formal tenure regularization at settlement level. Regularization policies are generally based on the delivery of individual freehold and, more rarely, leasehold titles. However, the difficulty of finding legal forms of regularization compatible with constitutional rules and the legal framework acceptable to the actors concerned, and in compliance with existing standards and procedures, constitutes a major obstacle for many operations.

The second approach emphasizes one of the components of formal tenure regularization policies, security of tenure. It does not require the provision of freehold individual titles, although this is not excluded. Rather it combines protective administrative or legal measures against forced evictions – including the provision of titles that can be upgraded, if required – with the provision of basic services. One of the objectives here is to preserve the cohesion of beneficiary communities and protect them against market pressures during and, more importantly, after the tenure upgrading process (Tribillon, 1995).

This approach must be understood as a first, but essential, step in an incremental process of tenure upgrading that can lead, at a later stage, to formal tenure regularization and the provision of real rights. Unlike complicated, expensive and time-consuming tenure regularization programmes, security of tenure can be provided through simple regulatory and normative measures (Adler, 1999).

The rapid integration of informal settlements through conventional tenure regularization and the provision of freehold titles may hinder community cohesion, dissolve social links and induce or accelerate segregation processes through market eviction. However, measures aiming primarily to guarantee security of tenure give communities time to consolidate their settlements, with a view to further improving their tenure status. Improvements to the economic condition of households, the emergence of a legitimate leadership at community level, the identification of right-holders, and the resolution of conflicts within the community and between the community and other actors involved (such as land owners, local authorities, planning authorities, central administrations in charge of land management and registration, etc), all form part of this consolidation process. In addition, the time between the decision to guarantee security and further formal tenure regularization and the delivery of property titles can be used to improve the quality of services in the settlement. It also gives households time to define a strategy and to save or raise funds to pay for the next step in the tenure upgrading and regularization process.

In addition, being given security of tenure without transferable or negotiable property titles lessens market pressures on the settlement and limits market evictions. This is an essential advantage of options emphasizing incremental regularization procedures, where occupants are granted occupancy rights that can, at a later stage, be incrementally upgraded to real rights, such as freehold or long-term leases, if so desired. Such an approach can be used both on vacant land and for regularizing irregular settlements (Christiensen and Hoejgaard, 1995; Fourie, 1999).

The shift from tenure regularization to tenure security policies

During the last decade in most developing cities the common perception has been that property titles are the best if not the only way to ensure security of tenure. Such approaches have achieved limited results. When large-scale allocation of property titles to households living in informal settlements has been made possible, it has often resulted in increased pressure from the formal property market within the settlement, and/or an increase in the cost of services, both of which have tended to exclude the poorest sections of the population. As suggested by Krueckeberg and Paulsen (Chapter 15), this calls for a critical analysis of the positive and negative consequences of increased formalization and commodification of the urban tenure process.

- Policies based on large-scale provision of land and housing by the public sector have proved to be ineffective in terms of reaching the poor.
- Market-oriented responses tend to increase social urban segregation as the formal private sector usually responds to the needs of households in the upper income bracket.
- Public–private partnerships in land and housing development cannot easily reach the poor unless heavy and well targeted subsidies can be provided (Payne, 1999a).

- Centralized land registration and management systems and procedures, and existing legal and regulatory frameworks cannot respond to the requirement of large-scale tenure regularization programmes in cities where up to 50 per cent of the urban population is living in irregular settlements.
- Governments rarely have sufficient human and financial resources to operate on a large scale.
- Shifting from projects to programmes and then to policies remains a major problem.

As stressed by Fourie (1999): 'Security of tenure policies should not be systematically linked with conventional approaches to tenure based on access to property titles. Security of tenure does not equal ownership.'

However, security of tenure is considered by many observers to be an inadequate response to the needs of households in irregular settlements when compared with tenure regularization (Urban Management Programme, 1989; USAID, 1991; World Bank, 1993; Ansari and von Einsiedel, 1998). More often than not, government officials in charge of land management believe that providing security of tenure tends to encourage the illegal occupation of land. For some this will result in a drastic reduction in their administrative power and hence in the related advantages they receive (including financial ones) through illicit practices. Landowners and informal land developers may feel that they have been deprived of their right to dispose of their property and have had no fair compensation, which they would normally expect to receive in cases of formal tenure regularization. For the communities concerned, to be provided with security of tenure may be seen as a necessary, though inadequate, measure that may postpone indefinitely their access to ownership. Security of tenure is not seen as a first step in an incremental tenure upgrading process, but rather as a 'temporary' solution, which can last forever if administrations do not have the capacity or the will to pursue the process. This is especially true in contexts where people do not have confidence in the promises and commitments of governments regarding tenure regularization. In this case, freehold is considered the only reliable and sustainable guarantee that they will not be evicted (Durand-Lasserve, 2000).

The choice between formal tenure regularization through access to land ownership and measures aimed primarily to guarantee security of tenure is the underlying debate in many chapters presented in this book.

Diversity of situations and objectives requires diversity of responses

Although there has been a considerable shift towards implementing more flexible forms of security of tenure, which tend to stress user rights rather than ownership (this is the case in India in particular), it has not produced, or has not yet produced, programmes and policies that can be applied at a national level.

The contributions in this book, especially the chapters on Brazil and South Africa, all underline the fact that conventional forms of legal regularization of irregular settlements are generally preferred, the ultimate goal being the issuing of freehold titles. However, they also describe the difficulties encountered in the implementation and especially in the enforcement of these methods. The contributions show the limitations of conventional responses, and also the developments that are currently taking place. As emphasized in the New Delhi Declaration (UNCHS, 1996c) and by the Habitat Agenda (UNCHS, 1999b), they highlight the clear need to have a variety of responses available in order to cope with the diversity of local situations encountered.

There may be various objectives behind the provision of security of tenure, such as ensuring social peace (the prime political motivation of most governments), social justice, urban planning or environmental and economic objectives like the integration of informal practices into the sphere of the formal economy.

As emphasized in the different chapters of this book, the content of security of tenure policies depends on the priorities given to these objectives and the forms and types of irregularity encountered. Clearly, the responses and options available to deal with security of tenure cannot be seen only in technical terms. They depend on a set of interrelated factors of social, political, economic and technical relevance:

- The principle of the right to housing and legal measures to enforce this right frequently contradict constitutional principles regarding the protection of property rights. This is one of the main areas of conflict when tenure upgrading and regularization policies are implemented, as well as when providing the simplest forms of secure tenure (Leckie, 1992, 1995).
- The respective responsibilities of central and local governments in relation to the implementation of security of tenure policies are generally clearly defined. More often than not, local entities have responsibilities regarding land and housing policies but are hindered in carrying them out by their limited resources, both human and financial (UNCHS, 2001, pp57–75).
- At the city/municipal level, the options available regarding security of tenure policies depend on the balance of power between various urban stakeholders as well as on the political orientation of the municipality.
- Available options also depend on the prevailing residential tenure systems in place.
- At the city level, another factor that has to be taken into account is the size of the population living in irregular settlements.
- At the settlement/community level, the measures employed will depend on the size of the community concerned, any political influence that may be involved, the age of the settlement and level of community organization, among other things. Any or all of these factors can determine whether the claims and demands of communities are in fact put forward for consideration.
- The role of NGOs and organizations of civil society must be considered in their local context (African NGOs Habitat II Caucus, 1996).

Responses also depend on the tenure status of the land occupied by irregular settlements. Land may be publicly owned: it may form part of the public domain of the state, in which case it cannot be alienated, or it may fall within the private domain of the state. Land may also belong to the central government or to a local entity (such as a state, province or municipality). In a large number of sub-Saharan African countries, for example, the land is owned and managed by the state, which supposedly, though this is not always the case, controls and regulates access to land through the discretionary allocation of occupancy permits, which can later be upgraded to leasehold and sometimes to freehold, and through tenure regularization projects (UNCHS, 1999c). However, achievements are limited, due mainly to the limited management and processing capacity of the central administrations involved, and to widespread corruption and illicit practices in the allocation process.

Land may also be privately owned, either collectively or individually. In this case, securing tenure will depend on negotiation with owners, the capacity of judicial power to make and enforce decisions, political will and the resources available for paying compensation through subsidies or cost recovery mechanisms.

The communal or customary system is, in many sub-Saharan African countries, the main provider of land for unauthorized developments. The success of tenure regularization policies depends on how the customary system in general is formally recognized, or whether it is simply tolerated or not even officially recognized at all.

Land belonging to religious organizations, in particular in Islamic countries or areas, covers a wide range of situations, but these lands are usually under rental tenure. In principle, religious land is protected against illegal occupations and from legislative encroachments, but it is not protected from market pressures, especially when located in prime urban areas. Conflicts between religious organizations or foundations that own or manage the land, and occupants, may turn the settlement into an irregular settlement. Then specific responses will have to be found in order to provide the populations concerned with some form of secure tenure.

Whatever the tenure status of the land, one factor plays an important role in the success or failure of security of tenure policies: actual or potential land-related conflicts. Whereas settling on disputed private land is often considered by poor communities as easier and less risky than settling on undisputed land, it may render the tenure regularization process extremely long and difficult.

Lastly, responses to the tenure regularization question also depend on the type of irregular settlement involved. Experience has shown that successfully securing tenure in squatter settlements will depend to a great extent on the legal status of the land (public or private land) and on the resources available for paying compensation if required, for providing basic services to occupants and for covering the costs of relocation of displaced populations whenever necessary (Durand-Lasserve and Clerc, 1996; Jones, 1998). Providing security of tenure in unauthorized land developments raises another series of problems, mainly relating to cost recovery for items such as formal land registration, titling and servicing. Informal rental, whether in squatter settlements or in unauthorized land developments, is the form of irregular occupation that raises by far the most difficult problems, as the populations concerned are poor and unorganized and are not necessarily seeking upgradable tenure arrangements.

A wide range of options are available to respond to the need for security of tenure (Fourie, 1999). These must be considered in the light of local circumstances and contexts:

- de facto recognition, but without legal status (this guarantees against displacement or ensures incorporation of the area into a special zone protected against evictions);
- recognition of security of tenure, but without any form of tenure regularization (the authorities certify that the settlement will not be removed);
- provision of temporary (renewable) occupancy permits;
- temporary non-transferable leases (India);
- long-term leases (may or may not be transferable);
- provision of legal tenure (leasehold or freehold).

These rights may be allocated on an individual or a collective basis, depending on local circumstances, and record systems and the allocation process may be centralized or decentralized. The population may or may not be involved in the process.

All the case studies presented here suggest that innovations regarding security of tenure policies must come mainly from new uses of existing tools, and this must be backed by the emergence of new initiatives on the part of civil society. Social processes need to be set in motion at settlement and city level, participatory processes need to emerge and new approaches to planning must be considered. The focus now needs to be on strategic planning rather than on 'inventing' new tools.

A FRAMEWORK FOR POSITIONING INDIAN, BRAZILIAN AND SOUTH AFRICAN TENURE SECURITY POLICIES IN THE INTERNATIONAL DEBATE

India, Brazil and South Africa offer a wide range of residential tenure systems and practices, ranging from regular (or formal and legal) to irregular (or informal and illegal). However, many are in-between, with some, but not all, having the attributes of legality. In order to lay the foundations for a comparative reading of Parts 1 to 3, a series of general, contextual issues are addressed in relation to the three countries.

Population size and urbanization

The first and most obvious comparative point to make about India, South Africa and Brazil is the enormous diversity in population size, standing at 1 billion, 40 million and 165 million, respectively.

India's political economy is based on agrarian concerns and it was only relatively recently that cities, urban land and urban poverty received much attention. The urban population was estimated at about 34 per cent in 2000. In 1991 it was approximately 26 per cent. A key determinant of urban population growth in India, as elsewhere, is migration from rural areas (Bhatnagar, 1996). Between 1951 and 1991 the proportion of the population living in cities with a population greater than 10,000 rose from 44.6 per cent to 65.2 per cent. Over the same period, the metropolitan share (those cities with a population of over 1 million) increased from 19 to 30 per cent (Mahadevia, 2000). It is estimated that India's 3697 urban centres (the 1991 estimate) will have increased to 5000 by 2021. Of these, approximately 23 are expected to have a population in excess of 1 million (ibid). The projected urban growth rate for 1991 to 2001 is almost 3 per cent per annum, similar to the 1981 to 1991 annual growth rate (Kundu, 2000).

By comparison, Brazil is intensively urbanized. Eighty per cent of the total population lives in cities and of this figure 40 per cent reside in the metropolitan centres (Fernandes, Chapter 6). Brazil's urbanization process was extremely quick and it occurred throughout the country (Souza, 1999). In 1940 Brazil was characterized by a mainly rural population; 31.2 per cent of the total population lived in urban areas at this time. In 1970 the urban population had grown to 56 per cent and by 1991 it stood at 75.5 per cent. Significantly, the urban growth rate, at nearly 2 per cent, was considerably slower in the 1980 to 1991 period than previously. In the 1970s and 1980s urban growth was concentrated in metropolitan areas, shifting to medium-sized cities and cities surrounding the metropolitan regions in the 1990s. A return to metropolitan urban growth, relative to that in medium-sized cities, may have been occurring since the mid- to late 1990s (Souza, 1999).

In South Africa, despite years of government intervention to prevent, then control, the urbanization of black South Africans, between 48 and 65 per cent of the population is urban. This amounts to between 19.6 and 26 million people. Cities and towns generate 80 per cent of the gross domestic product and it is projected that by 2020 some 75 per cent of the population will live and work in them (Republic of South Africa, 1995). South Africa's current urbanization rate is estimated at between 3 and 5 per cent per annum (ibid). South Africa's 'Urban Development Framework' (ibid) argues that continued urban growth can be anticipated and should be planned for, but that the growth rate is sufficiently normal to suggest that effective urban management is possible. While acknowledging the unpredictability of factors such as international migration, HIV Aids and an environmental crisis, it concludes that there is no justification for interventions that attempt to prevent, restrain or induce urban growth. Metropolitan migration is now lower than the demographic scenarios of the 1980s predicted, with an accompanying growth in secondary cities and large towns (Centre for Development and Enterprise, 1995). A noteworthy but unquantified aspect of urbanization in South Africa is the persistence of temporary, circular forms of migration.

Relevant powers and functions: their constitutional division and practical application

An understanding of the division and operation of intergovernmental powers and functions is an important basis for the three-country comparative analysis. Banerjee addresses the Indian institutional framework in her introduction to India (Chapter 2). She indicates that Indian state governments have discretion in the arena of urban development, housing and land, although in practice most state governments do follow central government directives and models in respect of law and policy, with appropriate adaptations. The national Five-Year Plans are important instruments for central government influence, as resources are made available to state governments for national priorities in the plans, which include shelter and basic services. The Housing and Urban Development Corporation and the National Housing Bank are important institutions in providing and regulating the flow of credit for housing-related matters. Banerjee highlights the importance of state government revenue departments that, as custodians of state land, undertake a range of land- and housing-related functions including registration of tenure and issuing of title deeds.

While housing is a state competence in India, national government has a stronger role to play in South Africa, as housing is a concurrent competence with provincial governments. Concurrency in respect of housing was a surprise outcome of South Africa's constitutional negotiations; expectations were high that it would be a national competence (Narsoo, 2000) in the context of an anti-federal African National Congress (ANC). In practice this means that national government makes policy and provinces implement it, national government allocates subsidies and provinces approve their disbursement. Provinces therefore have the power to control the types of housing projects that get approved and the kinds of developers (Baumann, 1998).

The 1988 constitution established housing as a concurrent power of the federal, federated state and municipal levels in Brazil. All three levels of government are to promote housing construction and the improvement of existing housing and basic sanitation, as well as combat poverty and marginalization and promote social integration (Fernandes quotes these passages in Chapter 6). However, local government has played a minimal role in the provision of housing options and public services to date. The federal government withdrew from direct involvement in housing delivery in the 1980s. Currently it allocates funds to state governments and municipalities for housing programmes (Engelbrecht et al, 1999). Federal withdrawal from housing responsibility in the 1980s, coupled with lack of funding, experience and capacity in local government, has resulted in inadequate official responses to the housing dimensions of the urbanization challenge, deepening the formal housing backlog in Brazil (ibid).

Municipal government was given constitutional status in India's 1992 constitution. Several states have devolved many of the municipal functions to local government, although decentralization is hampered by capacity constraints. These functions include urban planning, development planning, land use regulation, civic infrastructure provision, slum improvement and urban poverty alleviation.

In South Africa local government was accorded the status of being a third sphere of government in the new constitution (1996). National, provincial and local government are distinctive, interdependent and interrelated, although the role of local government is still evolving as it develops in practice over time (Republic of South Africa, 1998). South Africa's white paper on local government establishes the objective of developmental local government in terms of which municipalities play the central role in representing communities, protecting human rights and meeting basic needs (ibid). The standardized housing response in South Africa leaves little room for local alternatives, whether defined municipally or by urban informal communities. This institutional setup has its origins in the well-developed housing policy in place by 1994 (arising from the national housing forum process). Housing policy had a long lead-time (Narsoo, 2000), becoming the de facto driver of urban development in South Africa. Housing delivery has traditionally been the preserve of provincial government; provinces evaluate proposals, allocate subsidies, set minimum standards, monitor projects and administer the subsidy system (Engelbrecht et al, 1999). Due to municipal capacity constraints, provinces may remain key players in housing delivery, despite the policy of accreditation of local authorities to undertake housing projects. What the impact of developmental local government (particularly the formulation and implementation of local housing strategies and targets as part of the municipal planning process) will be on shaping local housing strategies and on their relations with civil society organizations in doing so, remain open questions.

In his introduction, Fernandes emphasizes the significance of the 1988 constitution in introducing several major changes in respect of local government in Brazil, including conferring on municipal authorities the power to pass laws governing the use and development of urban space and the power to implement urban policies. Private property rights are recognized as a basic principle of the economic order but the constitution states that urban property accomplishes its social function only when it attends to the fundamental requirements of 'city orderliness' expressed in the master plan. A master plan is an important development and urban expansion instrument affecting municipalities with populations over 20,000. The constitution also approved the right of special urban *usucapião* (adverse possession), under certain conditions, for those who have occupied private land. Despite the claim by jurists that these constitutional provisions are hampered by lack of regulation, a point also made by Fernandes in his introduction, progressive intervention policies have been articulated at the local level in Brazil, such as those in Partido dos Trabalhadores (Workers' Party) municipalities. Nevertheless, implementation remains a key challenge. For example, adverse possession procedures (*usucapião*) should, in principle, apply to 50 per cent of squatter settlements. In reality, only a small percentage of favelas have benefited. Local government is comparatively powerful in Brazil, prompting provincial government officials on a South African study tour to São Paulo and Rio de Janeiro to refer to the 'municipalization' of government functions (Engelbrecht et al, 1999).

Informal access to land and housing

The origins, perpetuation and functioning of the informal land market can be attributed to various exclusionary forces. For example, political exclusion applied in South Africa through racial discrimination and in Brazil through literacy criteria applying to the vote (Huchzermeyer, 2002). Despite the abandonment of legislated political exclusion and the advent of democracies, high degrees of socio-economic inequality remain. For example, the top 10 per cent of the population control approximately half of the country's wealth in both countries (Huchzermeyer, 2002). In India, socio-economic exclusion is evident in that 38 per cent of the population lives below the poverty line (Bhatnagar, 1996). Access to shelter through formal market systems remains unaffordable for a majority of the populations in India, Brazil and South Africa and is manifest in the proliferation of various forms of irregular access to land and housing.

About 55 million people live in Indian cities in very poor conditions of shelter and infrastructure, by informally occupying public land. It was estimated that approximately 22.5 per cent of the urban population were living in irregular settlements in 1994, in unauthorized subdivisions and squatter settlements, mostly on public land. Figures for most of the large cities are much higher: 55 per cent of the population of Mumbai, 40 per cent in Ahmedabad, 39 per cent in Pune, 30 per cent in Hyderabad and 22 per cent in Bangalore (Mahadevia, 2000). Dilapidated urban blight and slum areas account for about half of the population in megacities and approximately a quarter in other metropolitan cities (Bhatnagar, 1996).

Between 40 and 70 per cent of the population in Brazil's main cities live in irregular settlements (Fernandes, Chapter 6). In Brazil, the informally subdivided peripheral *loteamentos* and the more centrally located land invasions or favelas provide access to land to the urban poor, under conditions of varying tenure insecurity.

In South Africa, an estimated 18 per cent of all households (1.5 million households or 7.4 million people) live in squatter conditions, backyard shacks or in overcrowded conditions in existing formal housing in urban areas, with no formal tenure right over their accommodation. About 9 per cent of all households (780,000 households) live under traditional, informal, inferior and/or officially unrecognized forms of tenure, primarily in rural areas. Official estimates put approximately 58 per cent of all households in situations of secure tenure (Royston, Chapter 10).

Tenure status in main types of irregular settlements

A wide spectrum of tenure insecurity exists in the three countries, ranging from illegal occupation to minor violations in building by-laws. As Banerjee points out in Chapter 2, while the spectrum of irregularity is broad, not all forms suffer from tenure insecurity. This section identifies and compares the settlement types in which insecure tenure is prevalent in India, Brazil and South Africa, drawing on Chapters 2, 6 and 10 in which Banerjee, Fernandes and Royston introduce India, Brazil and South Africa, respectively. While categorizing assists in country comparisons, it should be noted that there is diversity within the types proposed, amply demonstrated in the country chapters in Parts 1 to 3. The three forms of settlement identified earlier, squatter settlements, irregular subdivisions and informal rental, are evident, to greater or lesser extents and with varying degrees of local specificity, in the three countries.

Indian squatter settlement occupants have no legal rights over land or its development, precluding them from access to services or obtaining building permission. Subdivision, the second settlement type, is illegal because it violates zoning and/or subdivision regulations or lacks the required permission for subdivision. This prevents building permission being acquired and precludes eligibility for access to infrastructure. People living in squatter settlements tend to be poorer than those in illegal subdivisions.

More centrally located *loteamentos* in Brazil are occupied by the upper and middle classes and are usually legal. In peripheral *loteamentos* the occupiers have bought their plots from whoever presented themselves as landowners. Regardless of frequent problems concerning the original ownership of the divided land, the new occupiers have signed contracts, paid for the plots and in most cases paid all due taxes; nevertheless they have often not been allowed to have their contracts registered at the registry office. Favela dwellers, at least at the time of the original occupation, lack any form of property or tenure title. In some cities, such as São Paulo, nearly 10 per cent of the population lives under informal rental arrangements, without any legal protection and in hazardous sanitary conditions, in overcrowded, dilapidated buildings in city centres.

In South Africa the phenomenon of shack farming is an example of irregular subdivision in which occupants rent land. Land allocation by 'shack lords', processes of land transaction characterized by violent conflict, and by traditional authorities, where there may be conflict over jurisdictions with local authorities, are further examples of this category. The term 'allocations' tends to be used in preference to 'subdivision', which is reserved for the formal and technical activity undertaken by professionals. In Brazil and India the plot is sold and occupants can provide a 'paper' of some sort; a deed of sale or a sale agreement sometimes authenticated by public authorities. This means that, when the owner of the land wants to take it back or when occupants do not pay rent, they can be formally considered as squatters (in the strict sense of the term). Shack farming as rental could be compared with the 'bustees' of Calcutta, which along with the 'refugee colonies' accommodate 42 per cent of the city's population.

Informal rental arrangements, especially subletting and sharing, are a comparatively significant form of shelter for the urban poor in South Africa. Backyard shacks are informal rental or subtenancy arrangements on sites mainly within former black group areas, especially African (as opposed to Coloured or Asian) townships. Relationships arising from overcrowding within existing township housing stock are examples of subletting and sharing. Tenure conditions in these examples are not as precarious as in squatter situations, as some kind of consent arrangement has been made.

Formal responses

It has been observed in India that over the five decades of planning there has been a distinct shift from overdependence and faith in the state to market-oriented responses in the 1980s. Liberalization and globalization of the economy, decentralization and increased participation of the private sector and communities in the planning process have been greatly emphasized. The cost recovery imperative of privatization requires greater community participation, which is in turn necessary for monitoring the quality of services provided by the private sector (Mahadevia, 2000). The national housing policy, approved by parliament in 1994, marks a turning point in Indian housing policy. It aims to promote a more equal distribution of land and houses in urban and rural areas, and to curb speculation in land and housing in harmony with macro-economic policies for efficient and equitable growth (Agrawala, 1996). It also promotes the withdrawal of the public sector from managing housing production, emphasizing an enabling or facilitating public role.

In Chapter 2 Banerjee summarizes the main Indian public sector responses as:

- land and shelter supply in public housing and sites and services projects;
- legitimization of irregular land occupation;
- improvement of basic infrastructure; and
- lowering of costs through provision of low cost credit for housing, relaxation of development controls and development of low cost building technologies.

Implementation of the housing programme is the most significant instrument in South Africa to address the delivery of secure tenure. Other initiatives, such as the privatization of public housing stock in former townships and attempts to address the formal system constraints to land development (most notably the Development Facilitation Act), are also important. In South Africa the housing subsidy scheme is the only instrument for tenure intervention. Predominantly greenfield in nature, the majority of projects result in the provision of individual ownership. The impact of the promising 'People's Housing Process', which promotes a people-driven approach to informal settlement upgrading, remains to be seen.

Brazil is characterized by the absence of any effective housing policy intervention to speak of until the mid-1980s. It is the actions taken by urban poor communities and some municipalities, with the support, in some cases, of NGOs and churches, that have had the largest impact on the options people have for access to land and housing. Although there are some public housing projects, the invasion of land (favelas) usually in central areas, followed by self-construction and the acquisition and division of plots in peripheral *loteamentos*, have been the principal alternatives for housing available to the urban poor. The most noteworthy formal response in Brazil has been major improvements to the legal order, especially the urban policy chapter in the 1988 constitution. The new framework for urban law has four characteristics that influence tenure security. It promotes popular participation in the planning process, it recognizes the social function of property, it accords authority to the master plan and it introduces new terms to adverse possession. However, a lack of regulation of the key progressive constitutional precepts by federal law limits their impact. The initiatives undertaken by several progressive cities have gone some way to formally address the land and housing conditions of the urban poor through, for example, the introduction and application of zones of special interest and participatory budgeting.

KEY THEMES FOR A COMPARATIVE READING OF THE COUNTRY CHAPTERS IN THE CONTEXT OF GLOBAL ISSUES

The chapters that follow illustrate the diversity between and within countries regarding the urban tenure situation and its characteristics. The following general themes are proposed as a 'route map' through this complex terrain. Despite the diversity, it is possible to identify cross-cutting themes, although their application in specific country contexts and responses to them might vary. These themes are developed and elaborated in the concluding part (Chapters 15 and 16).

The nature of tenure strategies for the urban poor

Protection against eviction

Following the international trend, protection against evictions is to be found in all three countries. However, market evictions continue to exclude the poor from access to land and shelter.

The forms of tenure being granted

While alternative forms of tenure are provided in India, individual ownership remains dominant in both Brazil and South Africa.

A shift from regularization to the granting of tenure security and an accompanying shift to temporary and incremental, rather than immediate, upgrading of tenure

These two global trends, discussed in earlier sections of this introduction, are more observable in India than in Brazil and South Africa, where preferences for regularization are evident.

The impact of the different forms of irregular settlement, especially informal rental

As in other countries, informal rental plays a significant role in housing provision for the urban poor in India, Brazil and South Africa.

Integrated approaches to tenure security for the urban poor, including the provision of basic services and increasing access to credit

A number of integrated development programmes are available in India, including the scheme for environmental improvement of urban slums, which provides for improvements to basic infrastructure in slum areas and implicit security of tenure. The urban basic services for the poor, with a focus on women and children, contained an integrated development package with an emphasis on health and socio-economic development through community organization and participation and convergence of other programmes, including tenure regularization. The national slum development programme provides for physical and social infrastructure and some assistance for shelter improvement. Although no mention of tenure regularization is made, there is a condition that improved slums will not be removed for at least ten years. A large number of direct and refinancing public sector institutions also exist, with a mandate to provide credit facilities to the poor at lower than market rates of interest and longer repayment periods (Banerjee, Chapter 2).

In South Africa housing plays a significant and leading role in the urban development agenda. Urban settlement intervention is narrowly focused on the provision of infrastructure and housing and funding is rigidly structured to match this. A standardized capital grant is allocated to defined categories of beneficiaries. This inflexibility in intervention means that integrated development is severely hampered. By comparison, in Brazil, the integration of informal settlements appears to drive intervention programmes. Project funding is flexible, without reference to beneficiary qualification criteria, an enormously divisive, complex and potentially corrupt aspect of South Africa's subsidy administration system. There is much more emphasis on aspects such as social infrastructure, enhancing transport access, street lighting, sanitation services and the numbering of properties (Engelbrecht et al, 1999). Housing is not the primary activity within any programme. It is only a priority where people need to move from unsafe conditions, and no funding is provided for the construction of houses, except for relocation (ibid).

The institutional framework and the range of actors involved

The role of local authorities in a context of decentralization

The three countries share a concern for decentralization, increasing the role of municipalities in governance and, to varying degrees, a problem of unfunded mandates. Decentralization is seldom matched with the resources required to operationalize it. Human and financial resource capacity constraints impact fundamentally on the realization of decentralization objectives.

The role of community-based organizations (CBOs), including partnerships with informal actors

Civil society mobilization in India in the 1980s, through the national campaign for housing rights, was instrumental in creating a political lobby for inclusive policy-making processes. NGOs, CBOs, trade unions, grass-roots movements and lawyers and professionals in the housing and urban development sector participated in consultations and awareness campaigns all over the country. Although the Bill of Housing Rights drafted by the campaign has not become law, space has been created for debate, dialogue, negotiation and improved flow of information.

In both Brazil and South Africa, social mobilization in the workplace and in the living space emerged strongly in the late 1970s and began to exert real pressure in the 1980s. In South Africa urban social movements (civics) organized around resistance to, and defence against, apartheid, of which housing and service delivery was a component. The emphasis was on dismantling apartheid through local struggles of 'ungovernability', which sought to spearhead the transition to democracy by making apartheid structures in townships unworkable. However, in the face of the standardized housing subsidy intervention in South Africa, there has been little evidence of articulation of alternatives to the uniform application of the housing framework since the advent of democracy, except for the 'people's housing' process. In Brazil, by contrast, the urban question is much more politicized, as poor urban communities organize themselves and lobby for support from local politicians. Progressive intervention policies by municipalities in individual towns and cities have been articulated in a context of greater political space and demands for alternatives (Huchzermeyer, 2002).

Decentralizing the registration system

In India registration is decentralized to state governments. The revenue department is the only authority vested with powers to undertake registration of transactions, tenure and issue of title deeds. Very often officers of the revenue department are placed in municipalities, urban development authorities and housing boards (Banerjee in Chapter 2). Brazil's system is highly decentralized with a number of locally based registries. Deeds registration is separate from cadastral systems, planning systems and valuation rolls, rating and taxation (Engelbrecht et al, 1999). The South African situation demonstrates a sophisticated cadastral system, deeds office infrastructure and local government rating systems, which benefits municipal management including land use management, land valuation, the levying of rates and taxes and local government income. However, the South African system is highly centralized; the deeds registries office is responsible for the registration of ownership rights and the office of the Surveyor General is responsible for the geographic database.

References

Abbott, J (1996) *Sharing the City: Community Participation in Urban Management,* Earthscan, London, p247

Adler, G (1999) 'Ownership is Not a Priority Among the Urban Poor', *Habitat Debate,* Vol 5, No 3, pp24–26

African NGOs Habitat II Caucus (1996) *Citizenship and Urban Development in Africa: Popular Cities for their Inhabitants,* Prepared for Habitat II, Istanbul, June

Agrawala, P (1996) 'Land Management within the National Housing Policy', *Shelter* (Special Issue on Land), HUDCO, January

Ansari, J H and von Einsiedel, N (eds) (1998) *Urban Land Management: Improving Policies and Practices in Developing Countries of Asia,* Urban Management Programme, United Nations Centre for Human Settlements and New Delhi and Calcutta, IBH Publishing, Oxford

Arandel, C and El Batran, M (1997) 'The Informal Housing Development Process in Egypt', *DPU Working Paper 82,* Development Planning Unit – The Bartlett, University College London

Audefroy, J (1994) 'Eviction Trends Worldwide – and the Role of Local Authorities in Implementing the Right to Housing', *Environment and Urbanization,* Vol 6, No 1, April, pp8–24

Aurejac, P and Cabannes, Y (1995) *Accompagnement et financement d'initiatives communautaires locales; Enseignements des expériences en Amérique latine et en Asie des opérations de construction ou d'amélioration de l'habitat menées à l'initiative ou avec le concours d'associations populaires de quartier,* GRET–PSH, Paris

Azuela, A (1995) 'La propriété, le logement et le droit', in *Régularisations de propriétés,* Les Annales de la Recherche Urbaine 66, mars, Plan Urbain, Ministère de l'Equipement, des Transports et du Tourisme, pp5–11

Baross, P and Van der Linden J (eds) (1990) *The Transformation of Land Supply Systems in Third World Cities,* Aldershot, Avebury, UK

Baumann, T (1998) *South Africa's Housing Policy – Construction of Development or Development of Construction?* Report to Homeless International

Benton, L (1994) 'Beyond Legal Pluralism: Towards a New Approach to Law in the Informal Sector', *Social and Legal Studies,* No 3, pp223–242

Bhatnagar, K K (1996) 'Extension of Security of Tenure to Urban Slum Population in India: Status and Trends', *Shelter* (Special Issue on Land), HUDCO, January

Centre for Development and Enterprise (1995) *Post-Apartheid Population and Income Trends: A New Analysis,* CDE Research Series, Johannesburg

Christiensen, S F and Hoejgaard, P D (1995) *How to Provide Secure Tenure for Informal Urban Settlers,* Ministry of Lands, Resettlement and Rehabilitation, Namibia

Christiensen, S F, Hoejgaard P D and Werner, W (1999) *Innovative Land Surveying and Land Registration in Namibia,* DPU-UCL Working paper 93, May

Cobbett, W (1999) 'Towards Securing Tenure for All', *Habitat Debate,* Vol 5, No 3, pp1–6

Cohen, M and Sheema, S (1992) 'The New Agendas', in N Harris (ed) *Cities in the 90s, The Challenge for Developing Countries,* Overseas Development Administration, Development Planning Unit, University College London Press, London, pp9–42

Dale, P (1997) 'Land Tenure and Cadastral Reform. New Opportunities, New Challenges', Paper presented at the iKusasa-Consas Conference on *Surveying Tomorrow's Opportunity,* International Federation of Surveyors (FIG), Durban, 24–28 August

Dowall, D E (1991) 'Less is More: The Benefits of Minimal Land Development Regulation', in *Regularizing the Informal Land Development Process,* 2: Discussion papers, US Agency for International Development. Office of Housing and Urban Programs, USAID, Washington, DC, pp9–20

Dowall, D E and Giles, C (1991) 'Urban Management and Land: A Framework for Reforming Urban Land Policies in Developing Countries', *Urban Management Programme Policy Paper 7,* UMP, Washington, DC

Durand-Lasserve, A (1993) *Conditions de mise en place des systèmes d'information foncière dans les villes d'Afrique Sub–Saharienne Francophone,* Programme de Gestion Urbaine Document de travail no 8, PNUD–Banque Mondiale

Durand-Lasserve, A (1994) 'Researching the Relationship between Economic Liberalization and Changes to Land Markets and Land Prices: The Case of Conakry, Guinea, 1985, 1991' in G Jones and P Ward (eds) *Methodology for Land and Housing Market Analysis,* University College Press, London, pp55–69

Durand-Lasserve, A (1998) 'Law and Urban Change in Developing Countries: Trends and Issues', in E Fernandes and A Varley (eds) *Illegal Cities: Law and Urban Change in Developing Countries,* Zed Books, London, pp233–257

Durand-Lasserve, A (2000) 'Security of Land Tenure for the Urban Poor in Developing Cities: Home Ownership Ideology v/s Efficiency and Equity', Paper presented at the global conference on the urban future, *Urban 21,* Berlin, 4–6 July

Durand-Lasserve, A assisted by Clerc, V (1996) *Regularization and Integration of Irregular Settlements: Lessons from Experience,* Urban Management Programme, Working Paper Series 6, UNDP/UNCHS/World Bank UMP, Nairobi

Engelbrecht, C, Chawane, H, Gollocher, R, van Rensburg, G, Mkhalali, X, van der Heever, P and Browne, C (1999) *Upgrading Informal Settlements: A Brazilian Experience,* Gauteng Department of Housing, 1 July

Farvacque, C and McAuslan, P (1992) *Reforming Urban Land Policies and Institutions in Developing Countries,* Urban Management Programme Policy Paper 5, The World Bank, Washington, DC

Feder, G and Nisho, A (1998) 'The Benefits of Land Registration and Titling: Economic and Social Perspectives', *Land Use Policy*, Vol 15, No 1, pp25–43

Fernandes, E (1996) *Law and Urban Change in Brazil,* Aldershot, Avebury, UK

Fischer, J E (1995) *Local Land Tenure and Natural Resource Management Systems in Guinea: Research Findings and Policy Options,* USAID–Guinea, Land Tenure Centre, University of Wisconsin, Madison

Fondation pour le Progrès de l'Homme (1992) *La réhabilitation des quartiers dégradés. Leçons de l'expérience internationale,* La Déclaration de Caracas, Dossier pour un débat, Délégation Interministérielle à la Ville, Novembre 1991

Fourie, C (1999) *Best Practices Analysis on Access to Land and Security of Tenure,* Research Study, United Nations Centre for Human Settlements (UNCHS), forthcoming UNCHS publication

Gilbert, A (1987) 'Latin America's Urban Poor: Shanty Dwellers or Renters of Rooms?' *Cities*, February, pp43–51

Gilbert, A and Varley, A (1990) 'Renting a Home in a Third World City: Choice or Constraint?' *International Journal of Urban and Regional Research*, Vol 14, No 1, pp89–108

Habitat Debate, United Nations Centre for Human Settlements (Habitat) (1997) *The Habitat II Land Initiative*, Vol 3, No 2

Hardoy, J and Satterthwaite, D (1989) *Squatter Citizen: Life in the Urban Third World,* Earthscan, London

Harris, N (ed) (1992) *Cities in the 90s: The Challenge for Developing Countries,* Overseas Development Administration, Development Planning Unit, University College London Press, London

Hibou, B (1998) 'Banque mondiale: les méfaits du cathéchisme économique', in *Politique Africaine*, 71, Les coopérations dans la nouvelle géopolitique, pp58–74

Huchzermeyer, M (2002) 'Informal Settlements: Production and Intervention in Twentieth Century Brazil and South Africa', *Latin American Perspectives*, Vol 29, No 1, pp83–105

Imparato, I and Ruster, J (2002) (forthcoming) *Participatory Urban Upgrading: Lessons from Latin America,* Cities Alliance Series, The World Bank, Washington, DC

International Federation of Surveyors (FIG) and United Nations Centre for Human Settlements (Habitat) (1997) 'Concluding Statement in Support of iKusasa–Consas Conference' on *Surveying Tomorrow's Opportunity*, Durban, 24–28 August

Jones, G (1998) 'Comparative Policy Perspectives on Urban Land Markets Reform in Latin America, Southern Africa and Eastern Europe', Workshop report, in *Land Use Policy* 16, The Lincoln Institute of Land Policy, Cambridge, Massachusetts, 7–9 July, pp57–59

Jones, G and Ward, P (eds) (1994) *Methodology for Land and Housing Market Analysis,* UCL Press, London

Kundu, A (2000) 'Urban Development, Infrastructure Financing and Emerging System of Governance in India: A Perspective', *Management of Social Transformation (MOST),* UNESCO, Discussion Paper 48

Laubé, E (1994) 'Le rôle du locatif privé dans l'habitat populaire. Pour un renouveau des politiques du logement (Peshawar, N'Djamena, Calcutta, Dhaka, Conakry, Abidjan)', Doctoral Thesis, Université de Paris VIII Saint-Denis, Institut Français d'Urbanisme, Paris

Leckie S (1992) *From Housing Needs to Housing Rights: An Analysis of the Right to Adequate Housing under International Human Rights Law,* International Institute for Environment and Development, Human Settlements Programme, London

Leckie, S (1995) *When Push Comes to Shove: Forced Evictions and Human Rights,* Habitat International Coalition

Mabogunje, A L (1992) *Perspective on Urban Land and Urban Management Policies in sub-Saharan Africa,* World Bank Technical Paper 196, World Bank, Washington, DC

Mahadevia, D (2000) 'Historical Note on Urban Policies', in *Urban Agenda in the New Millennium, India,* Report of a workshop held by the School of Planning, Centre for Environmental Planning and Technology, Ahmedabad, July 23–24

Mathey, K (1992) *Beyond Self-help Housing,* Mansell, London and New York

McAuslan, P (1998) 'Urbanization, Law and Development: A Record of Research', in E Fernandes and A Varley (eds) *Illegal Cities: Law and Urban Change in Developing Countries,* Zed Books, London, pp18–52

Mitlin, D (1997) 'Tenants: Addressing Needs, Increasing Options', Editor's introduction in *Environment and Urbanization*, Vol 9, No 2, pp3–15

Mosha, A C (1993) *An Evaluation of the Effectiveness of National Land Policies and Instruments in Improving Supply of and Access to Land for Human Settlement Development in Botswana,* Report prepared for the United Nations Centre for Human Settlements (Habitat)

Narsoo, M (2000) 'Critical Policy Issues in the Emerging Housing Debate', Paper presented at *Urban Futures Conference,* Johannesburg, July

Payne, G (1997) *Urban Land Tenure and Property Rights in Developing Countries: A Review,* Intermediate Technology Publications/Overseas Development Administration (ODA), London

Payne, G (ed) (1999a) *Making Common Ground: Public–Private Partnerships in Land for Housing,* Intermediate Technology Publications, London

Payne, G (1999b) 'Urban Land Tenure Policy Options: Titles or Rights?', Paper presented at the International Research Group on Law and Urban Space/Centre for Applied Legal Studies workshop on *Facing the Paradox: Redefining Property in the Age of Liberalization and Privatisation,* Johannesburg, 29–30 July

Pugh, C (1995) 'The Role of the World Bank in Housing', in C A Brian and S S Ravinder (eds) *Housing the Poor: Policy and Practice in Developing Countries*, Zed Books, London

Rakodi, C (1994) 'Property Markets in African Cities', Draft, mimeo, provided by the author

Rakodi, C (1999) *Economic Growth, Well-being and Governance in Africa's Towns and Cities: Policy Priorities and Challenges,* Department of City and Regional Planning, Cardiff University

Republic of South Africa (1995) *The Urban Development Strategy of the Government of National Unity,* a discussion document, Ministry in the Office of the President, Pretoria

Republic of South Africa (1998) *The White Paper on Local Government,* Department of Constitutional Development, Pretoria (www.local.gov.za)

Rochegude, A (1998) *Décentralisation, acteurs locaux et fonciers; mise en perspective juridique des textes sur la décentralisation et le foncier,* Ministère Délégué à la Coopération et à la Francophonie

Rolnick, R (1996) 'Urban Legislation and Informal Land Markets: The Perverse Link', Paper presented at the Department of Urban Studies and Planning, *MIT Informal Land Market Seminar,* November

Serageldin, M (1990) *Regularizing the Informal Land Development Process,* Working paper, Office of Housing and Urban Programmes, US Agency for International Development

Skinner, R J, Taylor, J L and Wegelin, A E (1987) *Shelter Upgrading for the Urban Poor: Evaluation of Third World Experience,* International Year of Shelter for the Homeless, United Nations Centre for Human Settlements (Habitat) and Institute for Housing Studies, Island Publishing House, Manila

Souza, C (1999) 'Politico-institutional aspects of Urban-environmental Management in a Transitional State: The Case of Brazil', Paper presented at the international conference *South to South: Urban-environmental Policies and Politics in Brazil and South Africa,* Institute of Commonwealth Studies, London, 25–26 March

Torstensson, J (1994) 'Property Rights and Economic Growth: An Empirical Study', *Kyklos,* Vol 47, No 2, pp231–247

Tribillon, J-F (1993) *Villes africaines. Nouveau manuel d'aménagement foncier,* La Défense, ADEF, Paris

Tribillon, J-F (1995) 'Contourner la propriété par l'équipement des villes africaines' in *Régularisations de propriétés,* Les Annales de la Recherche Urbaine, No 66, Plan Urbain, Ministère de l'Equipement, des Transports et du Tourisme, mars, pp118–123

UNCHS (United Nations Centre for Human Settlements) (Habitat) (1991) *Report on the Workshop on Land Registration and Land Information Systems,* UNCHS, Nairobi, 15–18 October

UNCHS (1993a) *Améliorer les systèmes d'information foncière et de reconnaissance des droits sur le sol dans les villes d'Afrique sub-saharienne francophone,* Série Gestion Foncière 3, UNCHS, Nairobi

UNCHS (1993b) *Support Measures to Promote Rental Housing for Low Income Groups,* UNCHS, Nairobi

UNCHS (1996a) *Access to Land and Security of Tenure as a Condition for Sustainable Shelter and Urban Development,* Preparation document for Habitat II Conference, New Delhi, India, 17–19 January

UNCHS (1996b) *The Habitat Agenda. Goals and Principles, Commitments and Global Plan for Action,* United Nations Conference on Human Settlements (Habitat II), UNCHS, Nairobi

UNCHS (1996c) *Access to Land and Security of Tenure as a Condition for Sustainable Shelter and Urban Development,* New Delhi Declaration, Global Conference, New Delhi, India, 17–19 January

UNCHS (1999a) *Implementing the Habitat Agenda: Adequate Shelter for All,* Global Campaign for Secure Tenure, UNCHS, Nairobi

UNCHS (1999b) *Urban Poverty in Africa: Selected Countries' Experiences,* Africa Forum on Urban Poverty, UNCHS, Nairobi

UNCHS and Burkina Faso Government (1999c) *Aménagement foncier urbain et gouvernance locale en Afrique sub-saharienne. Enjeux et opportunités après la Conférence Habitat II,* Rapport du Colloque régional des professionnels africains, Ouagadougou, Burkina Faso, 20–23 avril

UNCHS (2001) *Cities in a Globalizing World: Global Report on Human Settlements 2001*, Earthscan Publications, London

UNDP (United Nations Development Programme) (1991) *Cities, People and Poverty: Urban Development Cooperation for the 1990s,* A UNDP Strategy Paper, UNDP, New York

UNESCAP (United Nations Economic and Social Commission for Asia and the Pacific) (1990) *Case Studies on Metropolitan Fringe Development with Focus on Informal Land Subdivisions,* United Nations, Bangkok

United Nations Economic Commission for Africa (1998) 'An Integrated Geo-information (GIS) with Emphasis on Cadastre and Land Information Systems (LIS) for Decision-makers in Africa', Expert group meeting, Addis Ababa, Ethiopia, November

Urban Management Programme (1989) *Land Development Policies: Analysis and Synthesis Report,* Revised draft, UNCHS and UMP, Nairobi

Urban Management Programme, Nairobi, and Ministry of Foreign Affairs, France (1993) *Managing the Access of the Poor to Urban Land: New Approaches for Regularization Policies in the Developing Countries,* 1) Synthesis of main findings, 2) Summary of case studies, 3) Conclusion and recommendations of the Seminar, Mexico, February

Urban Management Programme, Nairobi, Ministry of Foreign Affairs, France, GTZ, Germany (1995a) *Urban Land Management, Regularization Policies and Local Development in Africa and the Arab States,* Abidjan, 21–24 March

Urban Management Programme, Nairobi, Ministry of Foreign Affairs, France, GTZ, Germany (1995b) 'Lessons from the Study and the Seminar', *Urban Land Management, Regularization Policies and Local Development in Africa and the Arab States,* Land Management Series 4, UNCHS, Abidjan, 21–24 March

USAID (United States Agency for International Development) (1991) *Regularizing the Informal Land Development Process,* vol 1, background paper, vol 2, discussion papers, Office of Housing and Urban Programs, USAID, Washington, DC

Vanderschueren, F, Wegelin, E and Vekwete K (1996) *Policy Programme Options for Urban Poverty Reduction,* Urban Management Programme, Policy Paper 20

Varley, A (1994) 'Clientelism or Technocracy? The Political Logic of Urban Land Regularization', in N Harvey (ed) *Mexico: The Dilemma of Transition,* University of London and IB Tauris, London

Varley, A (1999) 'New Models of Urban Land Regularization in Mexico: Decentralization and Democracy vs. Clientilism', Paper presented at N–aerus international workshop on *Concepts and Paradigms of Urban Management in the Context of Developing Countries,* Venice, 11–12 March, http://obelix.polito.it/forum/n–aerus

Von Einsiedel, N (1995) 'Improving Urban Land Management in Asia's Developing Countries', Paper presented at the United Nations Habitat II Regional Policy Consultation on Access to Land and Security of Tenure, Jakarta, 28–30 August, p19

Ward, P (1998) 'International Forum on Regularization and Land Markets', in *Land Lines,* Newsletter of the Lincoln Institute of Land Policy, Vol 10, No 4, pp1–4

Wegelin, E A and Borgman, K M (1995) 'Options for Municipal Interventions in Urban Poverty Alleviation', *Environment and Urbanization*, Vol 7, No 2, pp131–151

Williamson, I (1998) 'The Bogor Declaration for Cadastral Reform and the Global Workshop on Land Tenure and Cadastral Infrastructure to Support Sustainable Development', *Developing the Profession in a Developing World,* Federation of International Surveyors (FIG), Proceedings of the XXI International Congress, Brighton, UK, 19–25 July

World Bank (1991) *Urban Policy and Economic Development: An Agenda for the 1990s,* World Bank policy paper, World Bank, Washington, DC

World Bank (1993) *Housing: Enabling markets to Work with Technical Supplements,* World Bank policy paper, The World Bank, Washington, DC

World Bank (1999) *A Strategic View of Urban and Local Government Issues: Implications for the Bank,* Transportation, Water and Urban Development Department, Urban Development Division, Draft, World Bank, Washington, DC

Zevenbergen, J (1998) 'Is Title Registration Really the Panacea for Defective Land Administration in Developing Countries?', Proceedings of the International Conference on *Land Tenure in the Developing World, with a Focus on Southern Africa,* University of Cape Town, 27–28 January, pp570–580

Zimmermann, W (1998) 'Facing the Challenge of Implementing a New Land Policy. Lessons Learned in the Context of International Cooperation', Proceedings of the International Conference on *Land Tenure in the Developing World, with a Focus on Southern Africa,* University of Cape Town, 27–28 January, pp581–587

Part 1 India

Chapter 2

Security of Tenure in Indian Cities

Banashree Banerjee

AN OVERVIEW OF POLICY AND PRACTICE

The poor in Indian cities, as elsewhere in the developing world, are a disadvantaged group as far as legitimate access to land and land development rights is concerned. About 55 million people live on public land with no tenure rights and in very poor environmental conditions (GOI, 1991). To remedy the weak position of the poor, the government has undertaken a number of measures, including legitimization of irregular land occupation and improvement of basic infrastructure in informal settlements. Security of tenure, in particular, has been the subject of attention during the 1980s and 1990s, although experience shows that any form of state intervention contributes to improving the tenure situation of the poor.

Tenure regularization has found its way into policy via two different routes. The first is the functional approach, in which secure tenure is seen as a means of achieving definite objectives. These could be poverty alleviation, credit worthiness for housing loans, incentives for poor families to invest in shelter improvement or compensation for relocation of squatters and pavement dwellers. In contrast to this is the 'rights' approach, emphasizing that every citizen has the right to a secure place to live. The courts, activist groups and political manifestos have tended to follow this path. Whichever route is taken, the results are security of tenure in the form of legally valid tenure documents; or implied security of tenure brought about through public investment, notifications, court stay orders, political patronage or group solidarity.

Within this general picture, there are significant differences between cities as far as the security of tenure question is concerned. This is not only because of the different sizes of urban centres and the wide variety of growth dynamics, but also because of the differences in policies and practices in the different regions of the country. Ultimately the wider framework of urban development and governance of urban land plays a vital role in determining policy impact.

Governance of urban land in India

Functional domains of central and state governments

Under the federal system of government in India, urban development, housing and land are in the functional domain of the state governments. The central government can, and does, issue directives, provide advisory services and set up model legislation. However, it is up to the state governments to adopt policy and legislation in accordance with article 246 of the constitution. In most cases state policies and acts follow central government directives and models with appropriate modifications to suit specific local requirements. Examples are the Land Acquisition Act, Transfer of Property Act, Town Planning Act and Slum (Improvement and Clearance) Act (Banerjee, 1994; Shaw, 1996).

The central government's influence is also felt through conditions attached to resource allocations to state governments for thrust areas of national policy. Some of these areas are urban poverty alleviation, improvement of shelter and basic services for the poor and development of small and medium towns. The two national level parastatal finance institutions, Housing and Urban Development Corporation (HUDCO) and National Housing Bank (NHB), have devised credit packages in keeping with national policy. These have considerable impact on land assembly and delivery of housing and infrastructure by local and state level public institutions.

Municipal functions

The Constitution (74th Amendment) Act of 1992 aims at democratic decentralization and accords a constitutional status to municipal governments. As part of the Act, urban planning, regulation of land use, economic and social development planning, provision of civic infrastructure, slum improvement and urban poverty alleviation are among the functional areas for which powers are devolved by state legislatures to municipalities. However, municipalities have neither the required manpower nor the financial resources to take up planning and infrastructure provision. These continue to be looked after by state government departments (town planning, public health, engineering and public works, etc) and parastatal organizations (water and sewerage boards, housing boards and urban development authorities, etc) (Singh, 1996).

The revenue department

Of particular significance to the tenure issue is the revenue department of the state government, which administers urban and rural lands in a state. It is the custodian of all land owned by the state government and looks after the interests of the state government in all matters pertaining to land. It functions according to set procedures, rules and regulations as laid down in acts, ordinances, codes and government orders. The revenue department is the only authority vested with powers to undertake functions such as land expropriation, registration of transactions and tenure, and to deal with the issue of title deeds. The only land-related functions that the revenue department does not look after are land use planning and regulation. Officers of the revenue department are often placed in local and state government institutions involved in land use planning and regulation, land assembly, allocation, development and tenure regularization.

It is relevant to note that according to the Central Government Properties Act of 1948, state government laws do not apply to central government lands located in the states.

The courts

India has an independent judiciary. The courts are very often approached to deal with land related disputes. These disputes often relate to the following: the individual's fundamental right to property versus the power of the state to expropriate land for a public purpose (Supreme Court, 1996a and b); challenging the award in expropriation cases; and disputes about ownership and boundaries. 'Stay orders' and litigation are known to hold up development programmes for years, including housing projects for the poor. On the other hand, court stay orders have often prevented forced evictions and courts have passed judgements in favour of poor people under threat of eviction.

Security of Tenure in National Policy

In the case of India, there is no explicit national policy on tenure regularization. However, it is included in most urban and housing policy documents, sometimes being directly mentioned and at other times by implication.

Up to the 1970s, policy was based on the notion that slums and settlements with irregular tenure were a transient phenomenon. Slum improvement was considered as a temporary measure for ensuring healthy living conditions until residents were re-housed. The model Slum Areas (Improvement and Clearance) Act of 1956 provided the statutory basis and guidelines and was adopted by most states. Tenure regularization is not part of the Act but notification of an area as a slum under the Act, in itself, implies secure tenure, as residents cannot be evicted without the approval of the competent authority.

In the 1970s it was realized that affordable public housing was a distant dream and slum improvement was recognized as a long-term option. In 1972 a scheme for the 'environmental improvement of urban slums' (EIUS) was launched with joint central and state government funding to provide basic services in slums notified under the Slum Act. One of the criteria for the sanctioning of funds was a certificate from the municipality declaring that the slum would not be cleared for at least ten years after infrastructure improvements were undertaken. The purpose of this condition was to justify public expenditure, but it also led to some security of tenure for the residents. By 1996 about 40 million people had been covered under the scheme (GOI, 1997b).

The 1980s and 1990s stand out for the number of central government policy documents on urban issues. The 1983 report by the task force on housing and urban development was the first in a series of documents. It was extremely critical of the EIUS and the Slum Act for not paying sufficient attention to aspects such as tenure and land use. It recommended the integration of physical upgrading with the granting of legal tenure and with programmes for social and economic development, and the preparation of city level strategies for slum improvement. The task force recognized that deficient land supply was the root cause of irregular land occupation and recommended measures to increase the supply of land in cities (GOI, 1983; Banerjee, 1994).

A number of policy documents have followed the task force report and echoed its views. The Seventh Five Year Plan recommends, 'Steps should be taken to provide security of tenure to slum dwellers so that they can develop a stake in improving and maintaining their habitat' (GOI, 1985). The report by the national commission on urbanization (GOI, 1988) is another example. The India country paper for the 1996 Habitat–II conference in Istanbul stresses the importance of security of tenure for the poor (GOI, 1996b).

The national housing policy of 1994 considers an increase in the supply of land and access to basic services for all, particularly the poor, as crucial elements in realizing the policy objectives. The policy specifies,

> *The central and state governments must take steps to avoid forcible relocation or 'dis-housing' of slum dwellers. They must encourage in situ upgrading, slum renovation and progressive housing developments with conferment of occupancy rights wherever feasible. They must undertake selective relocation with community involvement only for clearance of sites which take priority in terms of public interest.*

The policy emphasizes the need to review land use controls and norms, as well as relevant legislation to enable the poor to gain access to land. A specific recommendation is to confer joint or exclusive land titles to poor and disadvantaged women (GOI, 1994).

The Eighth Five Year Plan (1992–1997) complemented the national housing policy and observed: 'Providing tenure rights is a precondition for success of any environmental improvement in slum areas.' The plan proposed the review of land-related legislation and slum acts to provide for conferment of occupancy rights. The Ninth Five Year Plan (1997–2002) further specifies the facilitator role of public institutions and identifies certain disadvantaged groups for direct land and shelter provision. It proposes incentives for the private sector to participate in the process of land and shelter supply for the poor.

The national housing and habitat policy (1998) of the present government is closely linked to the Ninth Five Year Plan (GOI, 1997a). It does not deviate from the principles of the 1994 national housing policy but places housing in the larger perspective of human settlements and the economy. One of the major concerns of the policy is to carve out a meaningful role for the private sector in the form of public–private partnerships. The policy calls for improved urban land information and adoption of innovative techniques like land pooling, land sharing, transfer of development rights, and land reservations for the poor to make land available in the market and for meeting the shelter needs of slum dwellers (Gupta, 1999).

The draft national slum policy (GOI, 1999) embodies the following core principle: 'Households in all urban informal settlements should have access to certain basic minimum services, irrespective of land tenure or occupancy status.' The participation of the community and the award of tenure documents are seen as necessary prerequisites to improvement of services.

The policy states:

> *Tenure shall be granted to all residents on all the tenable land sites owned or acquired by the government. Proper tenure rights shall be granted on resettlement or rehabilitation sites. Tenure shall be allotted in the name of the head of the household and spouse, except in cases where the head of a household is a single man or woman. Other forms of tenure may also be considered, if desired by the community. These may include group tenure, collective tenure, cooperative tenure etc* (ibid).

The range of plot sizes for granting tenure is to be jointly decided by the local body and the association of slum dwellers. It would be possible to transfer tenure rights by paying a fee and with prior approval of the residents' association. The policy proposes that the land use classification in upgraded areas should be 'high density mixed use'.

The draft policy outlines institutional roles for resolving issues regarding land owned by the central government. It is proposed that private land, on which tenable slums exist, is acquired by the local body (within a period of six months) and then individual tenure granted.

DETERMINANTS OF POLICY REORIENTATION

Policy changes of the 1980s and 1990s in favour of tenure security did not take place as a matter of course. Diverse, and even opposing, factors from within and outside the government machinery have converged and led to consensus on the tenure issue. These are briefly discussed below.

Political agenda

In the political economy based on agrarian concerns, extensive land reforms and tenure programmes have been implemented in the rural areas ever since the 1950s, whereas cities, urban land and urban poverty have received little attention. The enactment of the Urban Land (Ceiling and Regulation) Act (ULCRA) in 1976 was a departure from this change, with the agenda of redistributing surplus vacant land held by individuals for public use and low-income housing. Again, during the 1980s several state governments enacted laws to regularize the tenure of squatters on government land. In the parliamentary elections of 1989 and 1994–1995, provision of land and shelter for the urban poor and action against displacement were included for the first time in the election manifestos of most political parties. Such popular measures initiated through the political process are seen as 'zero cost intervention', where substantial benefits accrue to the poor at no or little cost to the public exchequer (Raj, 1990).

The role of civil society

Parallel to government policy formulation, the 1980s saw mobilization of civil society through the national campaign for housing rights. The campaign started in 1986 and gained momentum because of the following events: large-scale evictions in the cities of Delhi, Mumbai (formerly Bombay), Calcutta and Bangalore; large-scale loss of lives and livelihoods in precariously located areas because of a toxic gas leak in Bhopal; and proposed relocation of villagers from dam sites. Non-governmental organizations (NGOs), community-based organizations (CBOs), trade unions, grass-roots movements, lawyers and professionals in the housing and urban development sector participated in consultations and awareness campaigns all over the country. Central themes were the rights of citizens to a secure place to live and work and the prevention of displacement. The same themes have been the subject of public interest litigation filed by members of civil society against forced evictions. Starting with the Bombay pavement dwellers' case in 1984 (Supreme Court, 1986), court judgements have repeatedly taken the view that displacement is legitimate only when public interest is affected and only when it includes an acceptable relocation package (Banerjee, 1996).

The campaign questioned the contents of government policy and the 'closed' process of policy formulation. Even though the Bill of Housing Rights drafted by the campaign has not become law, it created a political lobby that demanded regional consultations on the national housing policy and the inclusion of 'non-officials' in the working groups set up for formulating the five-year plans. This has been accepted by the central government and applied to the housing policy, to the Eighth and Ninth Five Year Plans and to the currently ongoing process of national slum policy formulation. Thus space has been created for debate, dialogue, negotiation and improved flow of information.

The civil society movement has been further strengthened by the support provided by Habitat International Coalition (HIC) and the Asian Coalition for Housing Rights (ACHR).

International thinking

International thinking has influenced policy in India in two distinct ways. First, India, as an active member of the United Nations community, has undertaken activities recommended by UNCHS (United Nations Centre for Human Settlement), particularly in connection with the International Year for Shelter in 1987 and the two Habitat conferences in 1976 and 1996. The formulation of a national housing policy, an emphasis on the facilitator role of government, the recognition of the importance of the urban land issue and a positive view of irregular settlements are largely the results of imbibing current international thinking. The second way is through projects and programmes for poverty alleviation, slum improvement and sites and services funded by the World Bank,

UNICEF (United Nations Children's Fund) and bilateral agencies such as DFID (Department for International Development of the United Kingdom). Lessons from these programmes have helped to shape policy and practice in favour of slum improvement and tenure regularization.

Structural adjustment and market orientation

Recent policy documents project the market logic and structural adjustment policies to which India has been committed since 1993. Essential ingredients for this reorientation are institutional, fiscal and legislative reforms to enable the market to work. Included in this package is a concern for making the poor 'bankable', one of the ways being to provide land titles (Sundaram, 1995). Land expropriation is seen as an undesirable, anti-market practice, and for the same reason the ULCRA has been repealed by parliament. This narrows down the options for poor people to get legitimate access to land and implies that regularization of irregular settlements is acquiring even greater significance.

Lessons from experience

Both negative and positive experiences have contributed to the learning process. First is the realization that public housing options cannot fulfil the requirements of the growing mass of poor people in terms of cost, scale and variety (GOI, 1983). Second, irregular settlements are not a transient phenomenon and not enough attention has been given to the tenure issue in upgrading programmes. Third, land assembly at a large enough scale for housing projects and relocation is becoming extremely difficult and cannot be considered as a major option in future. In some states HUDCO could hardly fulfil its mandate of providing funding for low-income housing because of the difficulty in acquiring land for projects (Banerjee, 1996). Fourth, displacement and relocation have proved to be traumatic processes in cities such as Delhi, Calcutta, Mumbai and Bangalore and have led to even greater impoverishment and drastic political implications (Banerjee, 1994).

Then there are success stories of cases that have combined improvement of services with tenure regularization of slums. Some well-known examples are: the Calcutta *bustee* improvement programme; the Hyderabad community development programme; and the Mumbai and Madras slum upgrading programmes. Then there are the thousands of families who have benefited from statewide tenure regularization policies in Madhya Pradesh (Banerjee, 1994), Rajasthan, Gujarat and Andhra Pradesh. Through these learning experiences tenure regularization of irregular settlements has become a common practice instead of an exception.

INSTRUMENTS FOR POLICY IMPLEMENTATION

The central government has used three main instruments to support and implement its policies for making land, housing and services available to the poor: legislation, schemes and programmes funded by the central government and credit facilities. In fact, in the absence of explicit policies, the instruments themselves often provide policy directions.

Legislation

While the Slum Act provided the basis for improvement, clearance and relocation of slums, the Land Acquisition Act of 1894 has been the backbone for land assembly. However, in spite of amendments to make it more acceptable to landowners, its application in urban areas has become increasingly difficult. Disputes over ownership, boundaries, compensation and interpretation of 'public purpose' result in prolonged litigation. Moreover, the government's priorities have shifted to facilitating the private market to operate effectively, rather than building up public land reserves for housing the poor (Banerjee, 1996).

The ULCRA of 1976 was the other piece of legislation expected to make land available for the poor. Vacant land in excess of the ceiling limit was to be expropriated by the state and used for housing the poor and for provision of essential public services. On the whole, the measure met with limited success as the process got bogged down in litigation and inadequate institutional mechanisms. Finally, the ULCRA was repealed by parliament in 1998 to make it easier for private builders to operate.

There are a number of acts related to the acquisition of legal rights over land. These are the Transfer of Property Act of 1882, the Registration Act of 1908 and the Indian Stamp Act of 1899. Most state governments have exempted tenure regularization in poor settlements from the complicated and expensive procedures of these acts, initially on a case-by-case basis and later as a blanket provision.

Rent control acts of state governments were intended to safeguard the interests mainly of poor tenants against eviction and exploitative rent increases. But these measures have become a burden to owners, as properties do not give sufficient returns even for maintenance and repairs. This, in turn, has led to poor and dangerous housing conditions for the poor. A new rent control act came into force in Delhi in 1995 and several states are in the process of reviewing their acts.

Most often tenure regularization violates urban planning legislation in terms of land use zoning, subdivision and building regulations and standards. So far, urban development legislation has proved to be an impediment to securing full development rights in tenure regularization programmes. The Model Regional and Town Planning and Development Law, drafted by the Town and Country Planning Organization (TCPO) of the central government, provides guidelines for state governments to revise their legislation. However, it does not look into the tenure aspect at all.

Central government programmes

Starting with EIUS in 1972, a number of programmes and schemes for slum improvement and poverty alleviation have been launched under the five-year plans and implemented throughout the country. The usual practice has been to start with full central government funding and gradually transfer the responsibility to the state governments and municipalities. Except for the urban basic services for the poor (UBSP) scheme, which could include tenure regularization based on local demand, legal tenure is not part of the package. An undertaking by the municipality stating that there would be no relocation for the next ten years is considered sufficient to secure investment.

Credit facilities

There are a large number of direct and re-financing public sector institutions such as banks and housing finance institutions, all of which have a mandate to provide low interest credit facilities to lower income group families and cooperative societies. Legal security of land tenure is a necessary prerequisite for sanction of loans. This is with the exception of HUDCO, which also provides funds to public institutions on the strength of state government guarantees (GOI, 1996a). These institutions, however, pass on the loan to house builders with land mortgage as a security. This makes it impossible for residents of irregular settlements to obtain credit for house building. Further, the implementation of structural adjustment policies has placed limits on state government guarantees, making it difficult for public institutions to access loans from HUDCO for programmes such as housing the poor, municipal infrastructure improvement and land acquisition.

The NHB has taken a different approach to tenure in its slum rehabilitation scheme (SRS) initiated in 1993. One of the conditions of the scheme is that the implementing organization should grant tenure of the plot or of the dwelling unit to the beneficiary after redevelopment or reconstruction (Kundu, 1996). Funding is used as leverage for tenure conferment. In practice, the land invariably belongs to the state government and, unless there is a policy of granting tenure to settlers on government land, there is little chance of implementation.

Irregular settlements in Indian cities

Parallel to national policy, a variety of strategies, tools and techniques for tenure regularization have come into existence. Some of these pertain to an entire state, while others are city specific. They are devised and applied in different situations depending largely on the nature and scale of informality, the purpose of providing and/or obtaining secure tenure and the existing regulatory framework.

By far the most prevalent types of irregular settlements in Indian cities fall into the two broad categories of squatter settlements and illegal subdivisions. They occur on public and private land and also on village common lands and tribal customary lands in urban fringes.

Squatter settlements are included under the general term 'slum' but are also known under different names in different parts of the country: *jhuggi jhompri*, *jhopadpatti*, *hutment*, *basti*, etc. Land (public or private) is illegally occupied and building activity takes place regardless of, and/or in violation of, all development control regulations. Occupants have absolutely no legal rights over land or its development. In most cities such settlements are characterized by very poor shelter and infrastructure conditions and location on precarious sites. Illegal tenure excludes such settlements from getting building permission or access to regular city services. The poorest people live in these settlements (Banerjee, 1994).

Illegal subdivisions are known under names such as unauthorized colonies, unauthorized layouts, refugee colonies (West Bengal), slums, village extensions, etc. Land is subdivided illegally, usually by illegal developers, and sold as plots. The subdivision is illegal either because it violates zoning and/or subdivision regulations, or because the required permission for land subdivision has not been obtained. Land may be privately owned, under notification for expropriation, urban fringe agricultural land or common land of a village engulfed by city growth. The sale or transfer of land and hence ownership of the plot may have a legal or quasi-legal status, but because of the illegality of the subdivision, plot holders cannot get permission to build. In addition the area is not eligible for an extension of infrastructure services. The inhabitants are not as poor as those residing in squatter settlements and they have some means to make an initial investment on the plot (Banerjee, 1994).

As can be expected, there are major differences in policy and practice for tenure regularization in squatter settlements and illegal subdivisions. It is observed that strategies, tools and techniques are a by-product of the extent to which the state or local government is willing to tolerate and support squatter settlements and illegal subdivisions. The attitude of the authorities has also given birth to a range of interpretations and measures of security, which have a base other than legal tenure.

First the tools and techniques used to regularize squatter settlements are reviewed, followed by illegal subdivisions. Examples from some Indian states are given. The comparative effectiveness of different techniques can be measured in terms of the scale of operation, operational procedure, cost and bundle of rights made available.

Examples of strategies, tools and techniques for providing security of tenure to squatter settlements

Legal tenure

The extension of land tenure rights over government land, locally known as *patta*, to squatters is undertaken as a welfare measure. Tenure rights can be given in situ or in alternative locations on freehold, lease or licence basis. Even though there are cases of group tenure, granting of individual tenure is the general practice. Current approaches give preference to regularization in situ but relocation has been resorted to under specific circumstances. Often a combination of approaches has been followed (Bhatnagar, 1996). A number of states such as Andhra Pradesh, Madhya Pradesh, Orissa, Rajasthan and Maharashtra have opted for tenure regularization as a statewide policy across all urban areas. States such as West Bengal and Tamil Nadu have adopted a city-specific or programme-specific approach.

Between 1980 and 1998 about 9.5 million households in urban Andhra Pradesh have benefited from tenure regularization under the programme of allotment of house sites to the urban poor, administered by the revenue department. The Andhra Pradesh land revenue code and related government orders outline the procedures to be followed. According to the latest policy of 1995, those occupying state government land for more than five years are eligible for the issue of *pattas*. Poor families are given freehold *pattas* without any charges, while others have to pay the market price. *Pattas* are given in the name of women; they can be inherited but not alienated; and they can be mortgaged for obtaining housing loans. When tenure regularization is combined with housing, plots are invariably reconstituted.

Alternative plots with *pattas* are allotted to those living on sites that are 'objectionable' (environmentally unfit for habitation) or required for any essential public purpose. Squatters to be relocated are issued with identity cards to establish eligibility for the allotment of alternative plots. According to the rules, slum clearance schemes by municipalities are given preference in the allocation of government land.

For squatters or relocation on municipal land, the practice has been for municipalities to issue possession slips, which may later be replaced by revenue department *pattas*. For notified slums on private land, the government first acquires the land under the Andhra Pradesh Slum Improvement (Acquisition of Land) Act of 1956 and then gives individual *pattas*. When it has not been possible to acquire private land, land sharing has emerged as a successful alternative, particularly in the urban community development programme of Hyderabad, where CBOs are strong. Tenure regularization, together with housing and infrastructure improvement programmes, has benefited large sections of the poor in all urban centres. However, the tenure position of occupiers of central government land continues to be insecure.

In Madhya Pradesh state about 150,000 *pattas* have been granted to squatters on government land in the cities under the Patta Act of 1984 for the granting of leasehold tenure to landless people in urban areas. Under the Act any person occupying up to 50 square metres of land for residential use became eligible for non-transferable leasehold rights, either over the land occupied or in another location. The provisions of the Patta Act supersede zoning and subdivision regulations and building by-laws. Tenure regularization is implemented by the revenue department. Relocation to alternative sites was facilitated by the land reservation for the shelterless scheme implemented by the Madhya Pradesh government between 1982 and 1988. The scheme made it mandatory to reserve 5 per cent of the gross area of all residential projects – public, private and cooperative – for families without shelter. This land was handed over, free of cost, to the slum clearance board or the revenue department for allotment.

In the capital city of Bhopal, tenure regularization was combined with a municipal infrastructure programme and housing loans under the SRS of the National Housing Board, implemented by the Madhya Pradesh State Housing Board. About 20,000 poor households in Bhopal got lease documents. But the issue of *pattas* on private and disputed land and central government land remains unresolved, adversely affecting 8000 households (Banerjee, 1994).

In the eastern state of Orissa, the Orissa Government Land Settlement Act provides for the extension of tenure rights to the slum population on both a freehold and a licence basis. The type of tenure is decided on a case-by-case basis and on the strength of specific government orders. The Act lays restrictions on the transfer of plots and use for anything other than residential purposes. The Act is applicable throughout the state but has been implemented only in the five large cities and only partially, as settlements on private land and sensitive areas like canal banks are excluded from tenure regularization (Bhatnager, 1996).

Starting in 1971 with the state capital, Jaipur, Rajasthan adopted a systematic programme regularizing *katchi bastis* (squatter settlements) in all the 147 municipal towns in the state. In situ regularization is the preferred option. Ninety-nine year leases are offered with differential rates related to income. The process of regularization is seen as an ongoing activity of the revenue department with annual targets set by the state government. About 7 million people were included in the programme between 1982 and 1995 (Bhatnager, 1996).

The Maharashtra state government has used a mixture of legislation, government orders and innovative arrangements to grant tenure to squatters. The Maharashtra Slum Areas Act of 1971 provides for the protection against eviction of residents from areas declared as slums.

In Maharashtra state, tenure regularization was introduced in 1985, as part of the World Bank's slum upgrading programme (SUP) in Mumbai. Under the programme tenure was granted to registered cooperative societies of slum dwellers in an area. Granting of tenure to slums on state government land took place as soon as cooperatives were registered and there was agreement about the physical boundaries of the area. For slums on private land, the land had first to be acquired, under the provision of the Slum Act, but no progress was made. For central government land, a 'no objection' certificate was needed, but was not forthcoming, preventing about 50 per cent of Mumbai's slum dwellers from receiving the benefits of the programme. The SUP was extended by means of a resolution by the state legislature to all the municipal corporations in Maharashtra in 1985 but was not followed up.

In the 1990s the focus in Mumbai shifted to involve the private sector in providing housing with secure tenure to slum dwellers. In 1991 the state government initiated the SRS for slums on private land. This was essentially a land-sharing scheme in which the private landowner was encouraged to build apartments for slum dwellers who were resident on the land, using only part of the land.

The SRS was followed in 1995 by the slum redevelopment scheme (SRD), which is currently being implemented. Under the SRD all slum lands taken up for rehabilitation and owned by institutions of the state government can be leased out for 30 years at a nominal lease rate and increased floor space index (FSI) to developers. These developers will, in turn, build subsidized tenements for members of registered slum or pavement dwellers' cooperatives on part of the land. The crux of the programme is the use of land and development rights as leverage for making the private market provide housing for the poor. Modifications to the Slum Act, Maharashtra Regional and Town Planning Act of 1966 and Bombay Municipal Corporation Act of 1988 have been made to enable the participation of slum dwellers, NGOs, developers and landowners. However, developers are not attracted because of the fall in real estate prices (SELAVIP Newsletter, 1999).

Tamil Nadu does not have a statewide policy, but under the World Bank funded programmes – the Madras urban development project (MUDP) and Tamil Nadu urban development project (TNUDP) – *pattas* are issued on the upgrading of the slum and the recovery of the cost of development from beneficiaries. These *pattas* are in the form of lease-cum-sale deeds, which also permit home-based economic activities. About 150,000 households have been covered under these programmes.

West Bengal, like Tamil Nadu, does not have a statewide policy of conferring tenure rights to squatters. The Calcutta Thika Tenancy (Acquisition and Regulation) Act of 1981, amended in 1994, provides for the regularization of *bustees* or settlements on private land. *Bustees* are settlements with a distinctive three-tier arrangement; *bustee* dwellers or *bharatias* rent space in huts built by *thika* tenants on land leased to them by the landowners. The Act accords a legal status to *bharatias* and incorporates provisions against their eviction. Under the Act, land can be inherited but not alienated. Under the Premises Tenancy Act, *bharatias* are assured of receiving water supply and metered electricity connections.

Refugee colonies are the other kind of squatter settlements regularized in Calcutta and some other towns of West Bengal. These are peri-urban settlements on marshy or agricultural land, occupied largely by political refugees. Land occupation is in the form of organized squatting or purchasing use rights from the landowner. Most of these refugee colonies have been regularized in situ and provided with subsidized services.

In Delhi, the only form of secure tenure for squatters has been in relocation areas. The relocation of *jhuggi jhompris*, with various forms of tenure arrangements, has been implemented in Delhi since the 1960s and more than 230,000 plots/tenements have been allotted. Large-scale relocation has been possible because of public ownership and, therefore, availability of land. It is interesting to note that these relocation areas repeatedly violate development control rules and standards in terms of plot size, services and zoning. This has been justified by calling resettlement areas 'camping sites' and allotting plots with temporary tenure such as monthly licence or licence to occupy until asked to vacate. The fact that tenure is treated as anything but temporary was amply demonstrated by the poor response of residents to the offer of the administration to upgrade tenure to leasehold.

The 1990 'three pronged' strategy makes it possible to integrate relocation areas with other developments. Land reservations are made in all large projects for relocating squatters. *Jhuggi* dwellers who are members of registered multipurpose cooperatives become eligible for relocation plots, which are allotted to individuals as members of cooperatives. However, a judgement by the Supreme Court in 1998 to relocate 400,000 *jhuggi* families can hardly be feasible using this strategy.

Guarantee of continued occupation

There are some forms of official intervention that lead to security of tenure even without tenure regularization. An example is the undertaking from the municipality stating that a settlement will not be removed for ten years. Notification of an area as a 'slum' under the Slum Act amounts to some sort of security of tenure as it leads either to improvement or to relocation to an alternative site. In the case of Maharashtra, the Slum Act provides for protection against eviction from declared slums (Banerjee, 1996). States like Tamil Nadu and Karnataka do not, as a general policy, confer legal tenure to squatters but security of tenure is assured once public investment is made. In the case of Delhi, the latest policy is to encourage permanent buildings without granting tenure in squatter settlements – if the land is not earmarked for other projects.

At times court rulings guarantee continued occupation. In the Bombay pavement dwellers case (Supreme Court, 1986) and in the recent case in Delhi pertaining to 400,000 squatters, the ruling of no displacement without alternative accommodation leads to security because it is not easy to implement. Court stay orders also have a similar effect, though for a shorter time.

Other conditions leading to secure tenure

Settlements that have neither legal status nor government investment rely on other means to obtain security, or at least the perception of security. Settlements on land that is under litigation are known to remain undisturbed as long as the court case is not settled, sometimes for decades. Similarly, land that is not required for any other purpose is safe. However, very often such land is unsuitable for habitation or improvement. Examples are marshy or flood-prone lands, steep slopes and railway margins.

Political patronage from a local leader may lead to informal assurance against eviction and also provision of basic services.

There is a growing political lobby against eviction and media and judiciary support for the rights of squatters as citizens. Mumbai, Calcutta and Bangalore have strong CBOs, NGOs and grass-roots movements that have repeatedly confronted the government on evictions and tenure. This has forced the government to take positive policy measures, or at least to ensure immunity from displacement. Out of these confrontations has emerged a new trend of dialogue and negotiation where there is a willingness of all parties to work towards solving problems.

On a lower key, squatters have built up their own systems of security. Information through contacts with lower levels of bureaucracy gives early warning of impending demolitions. This prepares leaders to get stay orders from court, to mobilize demonstrations or to contact their political mentors. Politicians are invited for cultural or sports events; press reporters are invariably present and any statement on regularization is publicized. Of course, a price has to be paid for such protection and information.

Another common practice is to build religious structures in prominent places, knowing that the authorities are reluctant to demolish these for fear of communal problems, and hoping that surrounding areas will acquire immunity.

Any evidence of recognition by the authorities and proof of residence are considered important to establish a claim. Examples are ration cards of the public distribution system, identity cards, letters addressed to the family, tax receipts and electricity bills.

Examples follow of strategies, tools and techniques for providing security of tenure in illegal subdivisions.

Regularization

The official response to regularization of illegal subdivisions has tended to be city specific, unlike squatter settlements where state governments are invariably involved. Perhaps this difference exists because, in the case of squatters, regularization involves alienation of state government land. Except in Andhra Pradesh, where the state government has issued regularization guidelines for all the municipal corporations, it is the urban development authority or the municipal corporation that is responsible.

The process of regularization consists of either exempting the layout from subdivision and/or zoning regulations, or modifying the layout to conform to subdivision regulations, levying a regularization fee, payment for the extension of services and regularizing buildings on the basis of building regulations. Where applicable, the registration of titles is also required. There are local variations within this broad framework and some examples are discussed below.

Illegal subdivisions or unauthorized layouts exist, to some extent, in all the towns of Andhra Pradesh. The procedure to be followed in dealing with unauthorized layouts in Hyderabad, Visakhapatnam and Vijayawada is specified in a state government order issued in 1987. The urban development authority is first expected to take penal action against illegal developers or plot holders. If this is not possible, the layout is to be regularized by suitably modifying the layout and serving notices to plot holders to pay their share of regularization charges. These charges include the cost of services and roads along with external connections, the cost of community open space and layout approval charges. The procedure is only partially followed through because plot owners are willing to pay for services but not for layout regularization or property registration.

There are about 800 unauthorized layouts in Nagpur (Maharashtra) with 237,000 plots purchased from illegal developers by individuals or cooperative societies. The Nagpur Improvement Trust (NIT), as the planning and development agency for the city, is the implementing agency for regularization. The NIT has already modified the land use proposals in the draft development plan for Nagpur to accommodate 572 of the layouts that existed before notice was served for expropriation under the Land Acquisition Act and ULCRA.

However, on detailed examination, only 230 of these were found to be eligible for regularization. Residents are unwilling to pay the required rate for infrastructure provision as some form of arrangements for water and legal electrical connections was already available. The NIT has no proposal to regularize the remaining layouts (NIT, 1998).

A large number of unapproved layouts exist in scattered locations and around villages on the periphery of Chennai (Tamil Nadu). They have come up without any official sanction but land transfer has a quasi-legal status. Some of them are on excess land notified under ULCRA. The Chennai Metropolitan Development Authority (CMDA) regularizes layouts on the basis of a regularization plan and on payment of regularization and development charges.

An innovation here was the 'guided urban development' (GUD) component of the World Bank funded Tamil Nadu urban development project. The intention was for GUD to check unapproved layouts by creating competition in the supply of cheap plots. Small developers are encouraged to develop layouts in accordance with CMDA guidelines. The CMDA buys the land reserved for common facilities and roads and some plots for low-income families, while the developer can sell the remaining plots for housing and shops in the market. The lukewarm response to the scheme is attributed to lengthy procedures, lower profit margins and higher plot prices as compared with unapproved layouts, which continue to flourish at increasing distances from the city.

Illegal subdivisions in the form of 'revenue sites' emerged on agricultural land in the fast growing periphery of Bangalore (Karnataka) in the mid-1970s. The multiplicity of agencies and their overlapping jurisdictions in and around the city have created a laissez-faire situation in which it is easy for illegal developers to operate. The Bangalore Development Authority (BDA) has a policy for regularizing layouts on payment of fees, land use conversion charges (agricultural to non-agricultural) and development costs, particularly where plots have been purchased by cooperative societies as these can be exempted from the provisions of the ULCRA. However, the provision of services for scattered sites is hardly feasible.

Starting with 1962, there have been repeated official announcements about regularization of unauthorized colonies in Delhi, with advancing cut-off dates. At the same time the authorities have strongly condemned the practice and even carried out demolitions of offending structures. As Delhi is the national capital, the central government is also involved in these decisions. In fact the city institutions have repeatedly asked for central government resources for part of the cost of infrastructure improvement. The main problem facing the regularization process has been the unwillingness of plot holders to pay the entire regularization charges, as well as to follow the regularization plan. Despite almost four decades of regularization operations, only five colonies, out of about 800, have been fully regularized in terms of layout, lease deeds, services and facilities and payment of regularization and development charges.

Regularization of buildings

Cities in a number of states such as Andhra Pradesh, Tamil Nadu and Karnataka are conducting massive drives to regularize violations of building regulations on payment of regularization charges, as a way to build up municipal funds. In this process buildings in unauthorized layouts in a number of cities have also been regularized. Examples are Visakhapatnam and Hyderabad. Plot owners feel that with appropriate political lobbying, municipal services can now be obtained, even without paying for regularization or layout approval.

Declaration as slum

Sometimes illegal subdivisions are notified as slums under the Slum Act. They then become eligible for subsidized services and tenure becomes safe. Delhi, Vijayawada, Nagpur and Bhubaneshwar are cities in which this practice has been followed for a number of layouts. These areas become similar to site-and-services if plots are owned.

Announcement of regularization

The mere announcement of official policy or local government resolution to regularize settlements leads to immunity and lobbying for infrastructure provision. This has repeatedly been the case in Delhi and Nagpur. Irrespective of whether a settlement is regularized or not, inclusion of the settlement in the 'list' is projected as a guarantee for regularization. However, this need not be the case. Any kind of official survey or reconnaissance also leads to the assumption that regularization will follow.

Other conditions leading to secure tenure

The danger of demolitions disappears once layouts are built up. But in the initial stages the methods used to get a perception of secure tenure are similar to those used for squatter settlements. Religious structures are built, political patronage is sought and contacts with police and lower bureaucracy are maintained at a cost. All possible claims to legitimate residence are collected: ration cards, house tax receipts and electricity bills. Court stay orders against demolition are often obtained to buy time.

In towns where the scale of illegal subdivisions is small, or in small towns where most urban development takes place without permission, illegal subdivisions are not an issue at all and may not even be officially identified. Over time, houses are built, taxes collected and services provided.

Conclusion

There is no doubt that tenure regularization and related policies have led to the enhancement of security of tenure for a large population, particularly the poor, in Indian cities. They have legitimized the occupation of space in the city and provided access to development rights. This can be considered a significant achievement of the 1980s and 1990s, not only in terms of the results but also in terms of the change in attitudes towards informal settlements and the participation of civil society in decision-making. More open and innovative ways of approaching the land development issue have emerged. However, there is still much scope for improvement.

What happens to families who are not included and others who are yet to come?

This question points to the necessity of better mechanisms for negotiation and dialogue. It also argues for alternatives to the urban development paradigm that limits opportunities for the poor to gain access to land and rights to its development, leading to the creation of informality in the first place.

Which strategies, tools and techniques are effective and under what conditions?

State intervention with regard to regularization and shelter and infrastructure improvement measures has occurred most easily on land owned by the state and local governments. Understandably, it has faced more resistance on privately owned land, leading to innovative practices such as land sharing, SRD and land banks for the poor. Paradoxically, as much difficulty, if not more, has been encountered when settlements are located on land belonging to the central government, in spite of all its policy rhetoric.

An integrated approach that combines tenure security with improvement of services and house building and/or improvements to homes is more effective. Experience from states like Madhya Pradesh, Tamil Nadu and Andhra Pradesh shows that there are several ways of achieving this. In most cities there are limitations to the improvements that can be carried out in situ. Relocation to suitable locations through techniques such as land reservations for the poor, assignment of revenue land and SRD, in combination, seems to be more suitable.

There is no doubt about the superiority of legal security of tenure over the other options. The process of granting tenure is crucial. Linking tenure upgrading with cost recovery for infrastructure improvements, housing loans and reconstituting layouts, is more meaningful. On the other hand, 'lower' forms of security, ensuring basic improvement and protection against displacement without viable alternatives, are more suitable where tenure legalization cannot be taken up on a large enough scale. In any case, state intervention is a necessary condition to avoid exploitation and paternalism towards the poor.

Security of tenure is seen as part of the 'welfare' package in the case of squatter settlements. Limiting conditions such as time frame and use and transfer restrictions are usually imposed to ensure that the welfare measure is not 'misused'. More positive ways of achieving the objectives are seen in methods such as permitting home-based economic activities, granting tenure to women, or jointly linking tenure to cost recovery or shelter improvement and cooperative tenure. In other words, recognizing and providing for improvement of social and economic opportunities is important.

In relation to urban planning and development legislation, three approaches are distinguished: measures such as the Patta Act of Madhya Pradesh completely disregard all regulations; in Maharashtra and Tamil Nadu, the approach is to modify regulations to suit the new conditions; while in Andhra Pradesh the stress is on innovative use of existing laws and regulations. The first approach has problems in relation to overall city development objectives because it results in 'partial legality' and because of the limited opportunities for improving settlements in environmentally hazardous areas. A combination of modification and creative use of the regulatory framework is a pragmatic option. Greater attention needs to be given to the integration of regularization measures with city development. The incorporation of illegal subdivisions in land use zoning proposals in Nagpur is the first step towards that.

Effective strategies like land sharing in Hyderabad, land reservations for the poor in Madhya Pradesh and Tamil Nadu, GUD in Chennai and SRD in Mumbai are based on the creative use of the regulatory framework and an understanding of the land market. However, they cannot be seen as instant fixes. They need to be worked out properly and given time to take root. Radical policies like the socialization of land in Delhi or the ULCRA have good intentions and potential for making legal land supply accessible to the poor. However, they can only be effective when backed by political will, institutional mechanisms and procedures for implementation.

The participation of civil society in policy-making is a positive trend. Examples are national policy consultations and the formulation of relocation and rehabilitation guidelines in Mumbai. Even at settlement level, partnerships between local government, CBOs, NGOs and the private sector are emerging with positive results and towards more sustainable solutions. However, such relationships require time to take root and effort from all stakeholders.

Regularization is most often in response to a demand from the people most affected by informality – the people living in informal settlements. Community organizations such as cooperatives, neighbourhood associations and participatory approaches to planning and development lead to legitimization of demand making. This strengthens community solidarity and creates an awareness of rights and responsibilities. Such approaches are being advocated in policy and programmes and are practised in many cities, with local government and NGOs facilitating the process. The effectiveness of such measures in making government and donor interventions more effective is well documented.

What are the obstacles?

The regularization process is fraught with various kinds of conflicting interests between all the stakeholders: the conflict of interest between landowners and informal settlement dwellers; between central and state government; between environmentalist groups and regularizing authorities; between city planners for and against regularization; and conflicts within the local community. It is for this reason that public institutions implementing time-bound programmes often follow unilateral procedures that suppress or ignore these interests. However, the many cases where programmes could not be implemented, or where a long time is taken to arrive at settlements, show that these are a major impediment.

Short-term political gains have led to well worked-out programmes that do not fulfil the objectives. A case in point is the Patta Act of Madhya Pradesh in which cost recovery, plot reconstitution and environmental considerations were sacrificed for the sake of quick implementation. In addition, regularization practices often do not look into aspects such as the feasibility of infrastructure extension. Most often regularization is in violation of urban and land legislation and only partial solutions are possible.

Some of the conditions that allow informal settlements to survive are also obstacles to their regularization. Poor land information and poorly maintained property records lead to boundary and ownership disputes, holding up tenure conferment programmes and creating serious impediments. The multiplicity of land management institutions in a city – revenue department, municipality and development authority – also contributes to delays. Unrealistic development regulations and standards are often imposed on regularization of illegal subdivisions.

The lack of skilled manpower in sufficient numbers in local governments and revenue departments is a serious obstacle. This is particularly true in relation to innovative strategies requiring negotiation and working in partnership. Most often professionals see regularization programmes as politically motivated and not worth paying much attention to. The revenue department is hardly geared up to deal with urban land.

Most regularization programmes cannot be fully implemented because of poor cost recovery from beneficiaries. Very often costs of improvement are not calculated, keeping in view difficult ground conditions. In many cases the location of squatter settlements – on marshy or flood-prone land, steep slopes and highway margins – is an impediment to its improvement.

Has good practice filtered through?

Over the years a good interaction has developed between policy at state level and practice in cities. Good practice at city level has been scaled up into state policy, as in the case of Maharashtra and Andhra Pradesh. Central government policy and programmes also incorporate lessons from practice and advocate similar practices in other places.

There are examples of the transfer of techniques such as land sharing, land reservations for the poor and programmes such as urban community development, but the process needs to be made systematic. There is a tendency to transplant good practice without actually going into underlying conditions for success, often resulting in failures.

Programmes, apart from some donor funded ones, are not backed by evaluation and documentation. Official records and publications are more interested in projecting statistics and achievements. As a result process-related information is not available.

References

Banerjee, B (1994) 'Policies, Procedures and Techniques for Regularizing Irregular Settlements in Indian Cities: The Case of Delhi', *Urban Research Working Paper*, No 34, Vrije University, Amsterdam

Banerjee, B (1996) 'Urban Land: Is there Hope for the Poor', in Singh, Kulwant and Steinberg (eds) *Urban India in Crisis*, New Age International, New Delhi, pp359–370

Bhatnagar, K K (1996) 'Extension of Security of Tenure to Urban Slum Population in India: Status and Trends', *Shelter*, January, pp22–28

GOI (Government of India, Ministry of Works and Housing) (1983) *Report of the Task Force on Housing and Urban Development*, New Delhi

GOI (Government of India, Planning Commission) (1985) *Seventh Five Year Plan 1985–1990,* New Delhi

GOI (Government of India, Ministry of Urban Development) (1988) *Report of the National Commission on Urbanization*, New Delhi

GOI (Government of India, Planning Commission) (1991) *Eighth Five Year Plan 1991–1995*, New Delhi

GOI (Government of India, Ministry of Urban Development) (1994) *National Housing Policy*, New Delhi

GOI (Government of India, Ministry of Urban Affairs and Employment, National Building Organisation) (1996a) *Housing Finance Institutions Directory*, New Delhi

GOI (Government of India) (1996b) *India Country Report*, prepared for the Second United Nations Conference on Human Settlements, Habitat II, Istanbul

GOI (Government of India, Planning Commission) (1997a) *Ninth Five Year Plan 1997–2002,* New Delhi, Vol II

GOI (Government of India, Ministry of Urban Affairs and Employment) (1997b) *Annual Report 1996–1997*, New Delhi

GOI (Government of India, Ministry of Urban Affairs and Employment) (1999) *Draft National Slum Policy*, New Delhi, January

Gupta, J K (1999) 'National Housing and Habitat Policy 1998: Step in the Right Direction', *Shelter*, Vol 2, 2 April, pp22–25

Kundu, A (1996) 'Access of Urban Poor to Basic Services: The Changing Policy Perspective', in Singh, Kulwant and Steinberg (eds) *Urban India in Crisis*, New Age International, New Delhi, pp191–208

NIT (Nagpur Improvement Trust) (1998) *Report of Activities of NIT, 1997–1998*, mimeo, NIT, Nagpur

Raj, M (1990) 'Zero Cost Administrative Intervention in Urban Land Market', in Barros and van der Linden (eds) *The Transformation of Land Supply Systems in Third World Cities*, Aldershot, Avebury, pp267–275

SELAVIP Newsletter (1999) *Slum Redevelopment Schemes in Greater Mumbai*, April, pp67–72

Shaw, A (1996) 'Urban Policy in Post Independence India', *Economic and Political Weekly*, Vol 31, No 4, 27 January, pp224–228

Singh, K (1996) 'The Impact of Seventy Fourth Constitutional Amendment on Urban Management', in Singh, Kulwant and Steinberg (eds) *Urban India in Crisis*, New Age International, New Delhi, pp423–435

Sundaram, P S A (1995) 'Impact of Economic Reforms on Urban Land and Housing', *Shelter*, April, pp45–49

Supreme Court (1986) *Olga Tellis vs Bombay Municipal Corporation*, AIR 1986, Supreme Court 180

Supreme Court (1996a) *Chameli Singh vs State of Uttar Pradesh*, Supreme Court 549
Supreme Court (1996b) *State of Karnataka vs V Narasimhamurthy*, AIR 1996, Supreme Court 90

Chapter 3

Policies for Tenure Security in Delhi

Neelima Risbud

Delhi's experience of land supply has been unique. The informal land market has supplied substantial land despite the fact that formal land supply was supported by a strong land policy. This chapter examines the various tenure policies that have evolved in response to informal land delivery systems, the impact of such policies on improving settlement conditions, the causes of growth of such settlements and policy responses in terms of different types of tenure arrangements for improving security of tenure. Termed as 'substandard areas', these informal housing subsystems represent a parallel subsidized urban development process that, despite political patronage, is not integrated into the city. Each housing subsystem has undergone transformation over the past three decades. Policy responses have shifted with the changing situations. The context of urban land policy is described briefly below, followed by a discussion of the subsystems.

THE CITY DEVELOPMENT CONTEXT

Delhi has experienced a rapid growth in population since India's independence in 1947. The population of Delhi was 1.43 million in 1951, 8.5 million in 1991 and is expected to be 12.8 million by 2001. The spatial impact of such growth has been varied and poses a major challenge to city managers. A statutory master plan was devised for Delhi; a pioneering effort to manage the city.

The first master plan for Delhi was prepared under the Delhi Development Act as early as 1957 to achieve the planned growth of the city. However, the implementation of the master plan was not smooth. Major portions of low-income housing have developed over the past four decades in the form of squatter settlements – outside the master plan's legal framework for urban development. Even in the subsequent reviews, policies towards squatter settlements were never integrated into the master plan.

Delhi also has a complex administrative structure and the governance of the city is undertaken by various ministries of the union government, the state government, the Municipal Corporation of Delhi (MCD) and the Delhi Development Authority (the DDA under the control of central government) with its central and state legislators and municipal councillors. As Delhi is the national capital, all decisions related to land and city planning require the approval of the central ministry. As far as the land market is concerned, the influence of the city extends far beyond the administrative boundaries. The phenomenon of illegal land subdivision stretches into the agricultural green belt and also outside the union territory into other states.

URBAN LAND POLICY

Various factors, such as the increasing growth rate of the population, rising land values, growing speculative trends, overcrowding, lack of sanitation, traffic hazards, slums, substandard and unauthorized constructions, and the difficulty of implementing the master plan proposals on private land, led to the decision for 'large scale acquisition, development and disposal of land in Delhi' (Government of India, 1959). It was thought that the successful implementation of the master plan was possible only with the acquisition of bulk land. A decision was therefore taken to acquire all lands within urban limits and to keep them under the management and control of the DDA, creating government monopoly in land supply. This land was to be disposed of only on a leasehold basis. The scheme contemplated disposal of land to lower income categories for subsidized prices, while for higher income groups land was to be auctioned. Because of the subsidy involved, the lease conditions put a restriction on transfer for the first ten years. In the event of a sale, 50 per cent of the unearned increase of the value of the land would accrue to the DDA.

By December 1961, 30,800 hectares of land was notified under the Land Acquisition Act. By January 1986, 20,350 hectares were acquired by the Delhi administration and, out of this, 8750 hectares were actually given to the DDA for development. Possession of about 1420 hectares of acquired land could not be taken because of unauthorized structures on it (Ribeiro, 1987). This land became vulnerable to illegal use, constructions, etc. By 1992, the DDA had facilitated the construction of 1 million dwellings. These included 250,000 units on leasehold plots, 24,000 dwellings in resettlement of squatters, 23,000 leasehold apartments and 19,000 dwellings generated through leasing land to cooperative group housing societies. Despite this, large-scale squatting on public land and other informal land delivery systems have arisen and are discussed below.

SETTLEMENT TYPOLOGIES AND LAND TENURE

Informal land supply in Delhi has evolved and expanded more rapidly over the past three decades than the public supply systems. Private initiatives, which did not find a place in the official policy of land development, have arisen in informal settlements. Four major informal land delivery subsystems can be identified in Delhi: squatter settlements, resettlement colonies, unauthorized colonies and urban villages. These differ from each other in terms of process of development, tenure, the nature of illegality, the groups served and the resultant settlement characteristics. In 1983 more than 62 per cent of the population lived in various informal housing subsystems. These areas were not entitled to a standard level of services owing to their illegal/quasi-legal status.

Squatter settlements

Squatter settlements in Delhi are basically encroachments on public land by the poor. Land ownership and tenure is illegal but the extent of perceived security is varied. More recently, entrepreneurs have started promoting organized squatting.

In January 1990 a comprehensive survey was conducted by the food and civil supplies department via the issue of ration cards; about 929 settlements and 260,000 families were found squatting. Families squatting prior to the cut-off date of 31 January 1990 became 'eligible' for alternative leasehold plots. At present more than 600,000 families are squatting, out of which 70,000 families are located on land where priority public projects are to be implemented, 30,000 families are located on vulnerable/dangerous areas and 50,000 families are located on commercial sites. Therefore 450,000 families cannot be allowed to continue to stay at their present location.

Squatter settlements are in a very bad environmental condition, despite various government programmes over the past three decades to combat the problems. Acute deficiencies of basic services (for example, sanitation, waste collection and drainage) have made them susceptible to frequent epidemics of cholera, gastroenteritis, jaundice, typhoid, high infant mortality, and so on. Only regularization entitles the squatters to request a higher level of services.

Resettlement colonies

Programmes for resettling squatters, with varying tenure conditions and standards of development, have been implemented over the past four decades. The scale of this development has been massive. The first public policy towards squatter settlements was proclaimed in 1960 by the central government. A programme known as the '*jhuggi jhompri* removal scheme' involved the removal of squatters from public land and their resettlement in planned areas. Those squatting prior to 1960 were considered 'eligible' for alternative accommodation but those squatting after 1960 were to be evicted.

The scheme was revised in 1962 and 1967 because of the unabated increase of squatters and unauthorized sale of resettlement plots and it was decided to take up resettlement in a phased manner. In 1970 camping sites with minimum facilities were provided on a monthly rental (licence) basis for 'ineligible' squatters on the periphery of Delhi. Massive clearance and relocation was undertaken by the DDA between 1975 and 1977. The scheme did not conform to the land use stipulation of the master plan with regard to zoning regulation, plot sizes, building by-laws or municipal regulation of water supply and sewerage. The rental tenure of the plots enabled this deviation (Risbud, 1989). There was hardly any recovery of the licence fee for the plots.

The government recognized that although these plots were envisaged as camping sites, the colonies could not be treated as temporary. It therefore granted leasehold rights to residents and provided higher levels of amenities. During the next phase, after 1977, plot sizes for resettlement were increased. In 1983 built-up tenements were also rented as transit camps to families. So far, 1850 hectares of land have been developed for the resettling of squatters in Delhi. The emphasis shifted to the improvement of settlements on the same site.

After 1990 resettlement was taken up as a part of a 'three pronged' strategy for settlements in which land was required by a landowning agency for implementing a project. Resettlement was to be organized by setting up multipurpose cooperative societies of about 200 families each. The allotment of plots was on a licence basis through the cooperative society. So far 13,390 families have been resettled. Just before the state elections in 1998, the state government decided to offer leasehold rights to all the plot holders with licensed tenure in all 44 resettlement colonies. Plot holders were asked to apply and deposit a one-off payment towards ownership rights. This was in an effort to upgrade tenure and also to regularize illegal sales. However, the response was poor, as people did not see much benefit for themselves. At present the resettlement issue is under the consideration of the courts.

In order to provide land for resettlement, the government decided that the DDA would allocate 20 per cent of the land in all its urban development projects. For the first time resettlement was done in small pockets and integrated with the planned projects of the city. So far about 11,000 plots have been developed under this scheme. However, there is resistance from residents of adjoining areas to resettlements near their colonies.

The Delhi high court directed the administration that:

> *pending further orders, alternative plots for rehabilitation should be allotted only on a license basis with no right to transfer or part with possession thereof. Any transfer of possession would amount to automatic cancellation of the license without notice.*

Since the matter is still in court, no final decision has been made. In the meantime authorities allot camping sites in situations requiring priority clearance, and land for rehabilitation is made available with a different approach to resettlement.

Unauthorized colonies

The illegal subdivision of land in Delhi began in 1947. In 1962 there were 110 unauthorized colonies that housed a population of 221,000.

Formal land supply continues to be inadequate owing to delays in the process of land acquisition, development and disposal. The high standards of development adopted by the master plan have effectively made the

Table 3.1 *Land Tenure in Resettlement Colonies in Delhi (1960–1999)*

Year	*Programme*	*Tenure*	*Plot/ Size*	*Number of plots*	*Level of services*	*Rent/ Cost*	*Subsidy*	*Remarks*
1960	Plots Tenements	Leasehold	80 yd^2	3710 3767	Individual	2.5%	50%	Poor coverage, resale
1962	Phased shifting							Poor coverage
	i) Camping site	Rental	21 m^2		Shared	Rs3		
	ii) Regular plot	Leasehold	80 yd^2		Individual			
1967–1976	Camping sites	Monthly licence	21 m^2	197,624	Shared	Rs8		Massive relocation to periphery (political emergency)
1977	Regular plots	Leasehold	26 m^2	1722	Individual			Post emergency
1980	Built-up tenements Camping sites	Licence	10 m^2 12.5 m^2	1051 470	Shared			High cost, poor coverage
1990	Resettlement under three pronged strategy	Leasehold	18 m^2 +7 m^2 open	13,390	Individual		89% Rs 39,000	Resale of 50% plots Resettlement integrated with urban projects
1998	Tenure upgrading for licencee/ plot holder	Offer of leasehold rights				Rs5000		Plot holders perceive little benefit hence poor response
1998	Court decision to resettle 4,000,000 households	Licence			Shared			Non-availability of land for resettlement
1998	Proposed programme	Licence	15 m^2		Shared			For priority clearance

Source: Slum and JJ Wing, MCD (1998) *Growth of Slums/JJ Clusters in Delhi: Problems and Solutions*, compiled from various government reports

developments unaffordable to lower income groups. The difference in land price between the regularized and unauthorized developments is much higher than the associated costs borne by the residents.

A list of 1071 unauthorized colonies which developed prior to 1993 has been prepared. However, these have yet to be identified and verified from the aerial survey conducted in 1993. Many of these colonies are on private land, some are on government land and a few are on land notified for acquisition where compensation has already been announced and deposited by the DDA. Over a period of time the problem of unauthorized colonies has assumed large proportions and the residents have now become an important pressure group.

Urban villages

Urban villages in Delhi are those rural settlements that were engulfed by urban development. These settlements are organic and therefore not illegal. However, with the influence of market forces, these settlements expand and are informally transformed. The agricultural land belonging to these villages has been acquired for planned development by public agencies.

Rural settlements that have become part of urban areas have been declared as 'urban villages' under the Delhi Municipal Corporation Act of 1957. Under this provision 111 village settlements, to date, have been declared as 'urban'. Many unauthorized colonies have grown around the nucleus of urban villages. These urban villages house a population of 247,840 and cover an estimated area of 1607 hectares (DDA records, 1988). The growing demand for low-cost housing options in the city is being partly met by urban villages. There is a high demand for these properties because development controls and building by-laws, though applicable, are not explored. Most of the villages are on private land with the exception of seven on government land.

GOVERNMENT POLICIES

The Slum Areas Act

During the 1950s the problem of 'slums' in Delhi referred to dilapidated buildings and blighted areas of the old city. The Slum Improvement and Clearance Act of 1956 was enacted as a first step to deal with these slums, which were 'unfit for human habitation'. This definition only emphasized the physical aspect of a slum. It did not consider issues around informal tenure or the socio-economic conditions of the occupants. According to official estimates, about 2 million people are living in notified slums. These include squatter settlements, urban villages, unauthorized colonies and the entire old city.

The Act empowered the competent authority to ask the owners to carry out improvements or for the authority to carry out the improvements and recover the cost from the owners. It provided for clearance and redevelopment of structurally dangerous buildings and it provided for the protection of tenants against eviction. Over the past 40 years, the Act has been ineffective in improving the Old City, where extensive illegal conversions and rebuilding continue with total disregard for environmental considerations. The Slum Act was never meant for squatter settlements and it therefore does not recognize the considerable problems that exist because of variations of land tenure.

The improvement of squatter settlements

Under the 'three pronged' strategy adopted in 1990, improvement of squatter settlements was taken up in two ways. First, basic facilities under the environmental improvement scheme could be provided in squatter settlements where the land was not immediately required for a project but would be needed at a later date. Second, settlements where no projects were planned could be upgraded in situ with occupancy rights and civic services. These settlements would be allowed to remain and the dwellers would be helped to upgrade their shelters.

Improvement schemes

The 'environmental improvement of urban slums' (EIUS) was initiated by the central government in 1972 and continues even today. The scheme provides minimum services in notified slum areas through central grants. The facilities provided are street lighting, water hydrants or hand pumps, pay-and-use toilet complexes and dustbins. There is an expenditure ceiling of Rs800 per capita, which is inadequate. The Slum Wing of the municipal corporation implements the scheme.

The urban basic services programme (UBS) was undertaken by the partnership of the Indian government, UNICEF (United Nations Children's Fund) and the state government. The UBS was based on six guiding principles: community participation, child and mother focus, convergence, cost effectiveness, coverage and continuity. It aimed at providing an integrated package to improve health education and awareness using a participatory approach. Implementation of the programme started in Delhi during 1986 and 1987. The programme was revised and renamed urban basic services for the poor (UBSP) during 1990 and 1991. The objective of the scheme was to create community structures in terms of neighbourhood development committees. These would participate in the convergence and provision of physical, social and income-generating facilities. The scheme enabled the formation of registered societies of slum dwellers.

In an evaluation sponsored by the World Health Organization (WHO), Wishwakarma and Rakesh (1994) found that the implementation had been very poor. The UBS scheme was discontinued after the introduction of the UBSP. The budget provisions have remained unspent. The programme had serious management problems of intersectoral convergence and coordination. The objective of community organization and participation was on paper only. Delivery of services is therefore inefficient, piecemeal, duplicated and provisions are inadequate. Slum communities and non-governmental organizations (NGOs) were not represented on the committee, which was expected to steer improvements. The failure of such schemes is damaging as communities lose interest and faith in the government's commitment to create community structures.

Upgrading in situ involves the reorganization of slum dwellers on the same site with a planned layout, licensed cooperative tenure and provision of minimum basic services to enable families to construct their own houses. The Slum Wing implemented three projects on an experimental basis under this strategy and work on 4800 plots is in progress. Families were given 12.5-square metre plots on a licence fee basis. The shelter conditions and market value of these properties along with the cooperative/licensed tenure could not prevent unauthorized selling of plots to higher income families. The experience of implementing the pilot projects could not be utilized for wider application because of the reluctance on the part of landowning agencies to issue 'no objection' certificates for undertaking upgrading projects on their land. The high density of squatter settlements limits reorganization in a planned manner.

In addition to the above programmes, there have been a few political initiatives and support measures taken at various times by government. The Slum Wing has tried temporary solutions, for example, providing tankers for drinking water during summer and mobile toilets. These interim measures are too meagre to effect improvements and are very costly in the long run.

Perceived tenure security in improved slums seems to be short-lived and has been shaken in recent years by the legal intervention of the high court and the Supreme Court. The courts responded to a series of public interest petitions filed by middle-income flat owners against rapid encroachment of public land by growing slum clusters. It was agreed that only slum clusters officially recognized as of January 1990 were eligible for regularization and improvement. The courts wanted a clear policy on the relocation of ineligible squatters, as well as a plan of action to prevent new slums from forming. The city government expressed its inability to stop the growth of new slums in the absence of policing powers. It also highlighted the difficulties of land and resources in the relocation of squatters. In an attempt to make the city spotlessly clean, the Supreme Court ordered the removal of 400,000 huts and took the Delhi government to task. They asked city authorities to give serious consideration to the slum problem because if the situation in the city was not immediately addressed there would be no possibility of finding solutions in the future.

The authorities were faced with the problems of identifying and acquiring suitable land for relocation and mobilizing huge resources. After two decades the exercise of massive relocation is being reconsidered. The national alliance of people's movements has used the official admission that serious problems exist in the relocation policy, in part, as the basis for a new petition. This coalition demands basic civic amenities (water, sanitation and health for slum dwellers) as a right until an explicit policy is decided upon (Government of India, 1996[Q8]). Slum dwellers are organizing themselves and a slum dwellers' federation has recently been formed.

The health crisis in the form of frequent epidemics is a result of the fast deteriorating environment of the city. The most vulnerable pockets for epidemics are the approximately 1000 slum clusters in the capital, which lack basic amenities. The health perspective on slum improvement is becoming critical. The recent trend also indicates a class conflict between slum dwellers and adjoining higher income localities. According to *The Economic Times* (3 April 1991) the implementation of improvement schemes and the utilization of resources have been very poor and marked by inefficiency.

The Slum Wing is heavily dependent on NGOs for organizing and mobilizing communities. An NGO forum was initiated in Delhi in 1991 with a view to involving NGOs in slum improvement. There are more than 350 NGOs in Delhi but barely 10 per cent are involved in organizing slum communities. NGOs have been facing problems because of the absence of supportive policies for land tenure (Centre for NGOs, 1993).

The Slum Wing, which manages slum improvement, has been shunted several times between the DDA and the municipal corporation. Its present structure and mandate is hardly suitable to manage comprehensive sustainable slum improvements on a scale that will have a significant impact at city level. According to the 74th constitutional amendment, 'slum improvement' was one of the principal tasks of municipal authorities. Institutional capacity building is absolutely necessary to match the functional assignment of the Slum Wing.

Policies for unauthorized colonies

In 1961 it was decided to exempt built-up areas from statutory acquisition, under the 'large scale land acquisition and disposal' policy, and to regulate them with the following provisions: that they were built before the date of preliminary notification of the Land Acquisition Act; and that they could be fitted into the sanctioned regularization plan. This policy called for the regularization of 110 colonies. These colonies were regularized taking into consideration freehold tenure of plot owners.

In 1969 the government decided to also regularize (with leasehold rights) these colonies. However, the leasehold system could not be enforced because 62 colonies were located on non-conforming land uses (according to the master plan). The Delhi Lands (Restriction on Transfer) Act was adopted in 1972 to stop the sale and purchase of land; the sale of land in regularized colonies and urban villages without the permission of the competent authority was prohibited.

The central government announced another regularization policy in 1977 stating that all structures, within the union territory of Delhi, which arose through illegal subdivision of land up until June 1977, irrespective of their date of origin, were to be regularized (including those located in designated slum areas). Both residential and commercial structures were to be regularized once they conformed to a proper layout. Development charges, as determined by the authorities, were payable by the owners of the property for the provision of services. Those people displaced in the process of providing roads and community facilities were to be given alternative plots or accommodation. Land use as stipulated in the master plan was to be changed, wherever necessary, for the regularization of these colonies. Those colonies notified for acquisition were also to be considered for regularization; they were classified and dealt with by both the DDA and the MCD.

The residents considered the regularization charges too high. Out of the 607 colonies, regularization was started in 543 colonies, only five colonies were fully regularized and the others have remained at various stages of regularization. Recovery of development charges was poor; most people stopped payment after the first instalment. Anomalies between the actions of different institutions created enough ambiguity about the status of settlements to help them survive. Plots earmarked for facilities in the regularization plan were invariably encroached upon, or sold, by the colonizers. After two decades of regularization, the majority of areas lack basic municipal services (Banerjee, 1994).

The following land transfer mechanisms have been most commonly used: general power of attorney either registered or unregistered; receipt of amount against particular property on Rs2 stamp paper signed by both parties; and irrevocable will of landowner on Rs2 stamp paper, giving unlimited powers to the transferee to do anything with the plot of land.

The proposed regularization policy

In 1994 the Delhi government asked the members of the legislative assembly to submit plans of unauthorized colonies within their constituencies. This was followed by a newspaper advertisement requesting residents to hand in layouts of their colonies. Only a few layout plans were received. The city government recommended the regularization of 1073 colonies (which had arisen prior to 31 March 1994) to the union government. An aerial survey of the city carried out just before this formed the basis for the cut-off date.

A petition against the regularization move was filed in the High Court in mid-1993 by an NGO called Common Cause. It pleaded,

> *It was monstrous for the government to simply regularize the state once again, with no care for its effect on a city which is increasingly unable to cope satisfactorily with the basics of living.* (*The Times of India*, 1996, 10 December)

In December 1996 the Delhi high court bench gave an order to the union government to form 'a high powered committee' to decide within three months on every aspect of long pending issues of unauthorized colonies in Delhi.

In November 1997 the courts directed that until the matter of regularization was finalized, no further construction of any nature was to be undertaken in any of the unauthorized colonies of Delhi. The courts commented that the problem was reaching crisis proportions and made an analogy to that of war. In 1998 the court directed that the policy for regularization should be finalized, a clear definite cut-off date given and the amount set for development charges to be levied. The necessary penal action could then be taken against colonies which were not to be regularized; not just the demolition of a few selected houses but the colony as a whole. The courts referred to section 29 (Delhi Development Act) in their directives, which provides for imprisonment in cases of development without a layout. It further noted that the delay in decision-making was resulting in corruption at various levels. It was brought to the notice of the courts by the DDA that 'thousands of flats constructed by the DDA are lying unoccupied for want of electricity and water'. The court took notice of this and remarked that priority should be given to provide electricity and water to planned developments.

Ambiguities were noted on scrutiny of some colonies before regularization; many colonies had duplicate names and some colonies were not traceable. Boundaries of only a few colonies could be identified on available aerial photographs; for some only the location could be identified. Difficulties in physical identification of settlements are a serious hurdle to regularization.

An effort was made to shift the cut-off date and include the number and extent of colonies that appeared between 1993 and 31 December 1997. This could not be done owing to a change in government. There was a fundamental departure from the earlier policy in that colonies falling under the following categories were excluded:

- in notified or reserved forest areas;
- where the extension of physical and social infrastructure is not technically and financially feasible;
- where the majority of plot holders are not willing to pay development charges;
- where more than 50 per cent of plots are unoccupied and the number of plots is less than 400;
- where regularization would pose a serious hazard to public health, safety or convenience;
- which obstruct provision of infrastructure, proposed road width as per norms, railway lines, metro rail and other utility networks;
- where the built-up area exceeds prescriptions of the master plan; and
- where buildings have non-residential uses with areas of more than 50 square metres.

The development charges recoverable have been announced by government, which are lower than the actual estimates for provision of services. Politicians are demanding further reduction of development charges. The formation of a cooperative society of plot holders was a precondition for regularization to occur. Other regularization procedures were more or less similar to the earlier policy, except that in the new policy the society was given more responsibility. Despite the new provisions, the test of the policy will lie in its successful implementation. The new policy envisages the prevention of further growth of unauthorized colonies by prosecution of illegal colonizers by amending the Municipal Act and Delhi Development Act and by improving the vigilance on land.

Government policies towards urban villages

Until 1970 urban villages were ignored in the development process. Residential, industrial and commercial complexes were planned around the villages without physical integration. The DDA prepared separate development plans for these village settlements. The municipal corporation decided in 1963 that unless development work was completed in these villages, building permission was not to be approved in the extended areas of the village. In 1983 the government of India approved a scheme to provide basic amenities to 90 urban villages in Delhi. Funds were raised by the DDA and supplemented by central grants. The village beneficiaries were expected to pay for individual water supply, electricity and sewerage connections. The areas earmarked for community facilities, parks and road widening in the development plan of the village were to be acquired. Many of the properties on the periphery of the villages were notified lands under the Land Acquisition Act. Possession could not be taken of these properties as they had partly built structures on them. The Land Transfer Act of 1972 also covered these properties, whereby their sale was not permitted without prior permission. However, most of the properties were informally changing hands on 'power of attorney'.

A concession was given by the Delhi municipal corporation in 1959 to permit the setting up of cottage industries and to allow power connections. The result has been the indiscriminate and haphazard growth of industries in the villages. In 1983 an effort was made to restrict these industries by granting power connections to village residents only and by stopping the setting up of obnoxious and hazardous industries. In practice these policies have not been effective. However, the situation seems to have improved in terms of basic services, and agencies claim that 80 per cent of the development work is complete.

Other policy measures

A major shift in land policy is that the union government, in principle, has decided to allow private colonizers and builders to supply land and housing in Delhi. With this the era of public sector monopoly of land will come to an end, but whether it will reduce informal supply remains to be seen. Recognizing the futility of restricting land transfers, permission is being granted to leasehold property owners to convert to freehold properties on payment of a certain amount. The government has gone a step further and is recognizing properties purchased on power of attorney. A decision was taken to regularize unauthorized construction in legally approved buildings. All these regularization measures appear to be a *fait accompli*.

However, with the change of the central minister (2000–2001), the central ministry had reversed its liberal stand and started a major campaign against encroachment and unauthorized construction. Eviction of squatters and resettlement is being vigorously pursued; it was planned that in 2001 30,000 squatters were to be shifted despite strong opposition by politicians from all parties. The 'five pronged' strategy to combat violations includes detection, demolition, prosecution, cancellation of lease and eviction. Similarly, in the event of any transfer of plots in the resettlement colonies, the properties are sealed and the buyers evicted. Hundreds of families have been evicted under this campaign. The recent changes highlight the issues of sustainability of government policies and the process of decision-making.

CONCLUSION

To correct distortions in the land market, Delhi adopted a policy of socialization of land and government monopoly of assembling and disposing of land. Land transfers were discouraged through the enactment of laws and stringent tenure conditions. The failure of authorities to supply enough land led to the manifestation of informal channels by private sector initiatives; that is, squatters, unauthorized colonies and illegal land transfers of legal properties on power of attorney. Encouraged by subsequent regularization and the absence of any deterrent action, the informal settlements have grown rapidly. More than 60 per cent of the population lives in substandard settlements with quasi-legal land tenure. Several lessons were learnt during the four decades of implementation, and have prompted corrective measures. Informal settlements have been recognized and policies of regularization have been adopted. Various policy announcements on regularization have given various degrees of tenure security to different informal settlements. Regularization measures include legalizing illegal land subdivisions, granting legal tenure to squatters, regularizing land transfers on power of attorney, allowing the conversion of leasehold properties into freehold ones, regularizing unauthorized construction in formal developments and extending infrastructure to squatter settlements. The recent policy initiative of allowing private sector participation could have had a major impact on land policy, but the minister had retracted it during 2000–2001. With the new minister, the decision is being reversed and 'land pooling' is being considered.

To arrest the fast deteriorating environment that is endangering the city's health, minimum physical infrastructure is provided for all these settlements on 'humanitarian' grounds, irrespective of their tenure. Unauthorized colonies that are not regularized also manage to get some infrastructure improvement through political patronage. Squatter settlements are provided with fully subsidized minimum infrastructure under the environmental improvement scheme, thus giving them perceived security. However, squatter settlements are vulnerable to evictions when pressurized by landowning agencies to recover land, and when higher income residents of surrounding areas move the courts for their removal.

In situ tenure regularization has been limited because of resistance from landowning agencies to part with land. Tenure for squatters has been legalized mostly by resettlement. The large-scale resettlement approach is unique in terms of scale but cannot be a solution for squatters because, at the present rate, it would take a few decades to resettle all squatters. The slum wing is highly dependent on the DDA for making land available for resettlement, which is a major constraint for wider coverage. Moreover, the result of the resettlement process is a city with an unbalanced social structure with a periphery dominated by the poor.

The recognition of informal settlements has required changes in planning standards, which are now more responsive to the needs of the poor. The reduction of minimum plot sizes from 67 to 15 square metres in resettlement, and a consequent increase in density, is a case in point. Intermediate tenure 'right of occupancy' both in case of in situ upgrading or resettlement enables some improvement. But access to individual services and formal credit is denied. Violation of all conditions of regularization can be seen in regularized squatter settlements and resettlement areas. Licensed tenure or cooperative tenure has not been able to prevent the resale of plots. Offers by government to upgrade tenure from licence to leasehold have drawn little response because plot holders perceive little benefit.

However, the regularization of unauthorized colonies has resulted in eligibility for higher levels of infrastructure and legally recognized tenure but compliance with conditions of regularization is almost absent. Decisions on regularization have been politically initiated and are approved by central government, while the implementation rests with local bodies. Cut-off dates – which are subsequently shifted – have been the basis for deciding on regularization.

All these areas receive subsidized services which further burden the strained city infrastructure. Since these areas represent a major vote bank, infrastructure is provided through political concerns and sometimes is given priority over planned formal developments. The city has to bear the entire cost of servicing informal developments, both in terms of services and social facilities.

Regularization has resulted in some physical improvements and a substantial increase in property prices. In the case of unauthorized colonies, when prices become unaffordable even for middle class families, the process of plot subdivision starts so that plots become smaller and more affordable (Mitra, 1983). Good shelter consolidation and development of rental housing can be seen in most settlements. Proliferation of industrial and commercial activities in regularized unauthorized colonies and urban villages is very high (sometimes 70 per cent) resulting in environmental degradation. Regularization is perceived to give impunity from all laws of urban development; thus, these are attractive areas for unlicensed enterprises.

Substandard services and poor maintenance have resulted in serious environmental problems, which have attracted judicial intervention. This judicial intervention is indicative of the growing awareness and aggressiveness of environmental NGOs. Regularization has come to mean gaining legitimacy without substantially improving environmental conditions. In the light of this, the need to examine the impact of regularization on the city before granting legitimacy to double the number of unauthorized colonies is justified. Another significant trend is that local bodies are unofficially giving priority to the provision of infrastructure to unauthorized colonies, at the insistence of local councillors, and denying it to planned development. This is already causing resentment among residents of planned neighbourhoods.

In principle there is a greater recognition on the part of government to involve people (squatters as well as unauthorized colonies) in the process of regularization. However, the decision-making is very much centralized with politicians, and now with the judiciary, without involvement of stakeholders. Frequent political changes lead to the reversal of decisions that affect sustainability.

Major problems exist in the implementation of policies, especially in enforcing regulatory provisions. Recovery of dues for development inputs is extremely poor. Centralized decision-making, multiplicity of authorities, weak governance and exploitation of informal systems, both by politicians and market forces, have created major distortions. The proportion of unregistered land is rapidly growing. Management of informal tenure and financing of improvements is a serious challenge and cannot be met without a strong political will and an efficient administration.

References and Bibliography

Anthony, J (1991) 'Patterns and Processes of Illegal Land Subdivision, Delhi', unpublished thesis, Housing Department, School of Planning and Architecture, New Delhi

Banerjee, B (1994) 'Policies, Procedures and Techniques for Regularizing Irregular Settlements: The Case of Delhi', *Urban Research Working Paper 34*, Institute of Cultural Anthropology, Amsterdam

Centre for NGOs (1993) *Sustaining Slum Improvement*, Proceedings of the Workshop, 21 September, New Delhi

Chitra, S (1991) 'Housing Transformation in Resettlement Colonies, Delhi', unpublished postgraduate thesis, Housing Department, School of Planning and Architecture, New Delhi

Dube, N (1996) 'Urban Village: Concept and Reality', unpublished undergraduate thesis, Department of Physical Planning, School of Planning and Architecture, New Delhi

Government of India (1959) unpublished report of the committee set up to study 'the problem of introducing measures of control on land values and stabilizing land prices in urban areas of Delhi'

Government of India (1996) *India National Report for Habitat II*, Ministry of Urban Affairs and Employment, New Delhi

Gupta, A (1991) 'Changing Profile of Unauthorized Colonies with Reference to Production Activities, Delhi', unpublished postgraduate thesis, Housing Department, School of Planning and Architecture, New Delhi

Mitra, B (1983) 'Substandard Commercial Residential Subdivision in relation to Housing options in Delhi', report No 966, Institute of Housing Studies (IHS), Rotterdam

Mudassir, S M (1991) 'Impact of Infrastructure on Health in Squatter Settlements: Case Study: Delhi', unpublished postgraduate thesis, Housing Department, School of Planning and Architecture, New Delhi

Ribeiro, E F N (1987) 'Urban Land Policies in Context of Development Plan for Delhi', unpublished paper presented at NIUA/AIT regional seminar on managing urban land development, National Institute of Urban Affairs, New Delhi

Risbud, N (1986) 'Socio Physical Evolution of Popular Settlements and Government Supports: Case Study of Bhopal', *IHSP Research Report No 3*, HSMI, New Delhi

Risbud, N (1989) 'Relocation Programme of Rental Housing in Delhi', in *United Nations Centre for Human Settlement (Habitat) 'Rental Housing*, Proceedings of an Expert Group Meeting

Risbud, N (1991) 'Informal Land Servicing in Delhi', Report prepared by Mehta and Meera in *Land for Shelter: Delivery of Serviced Land in Metropolitan India,* vol 2, USAID, New Delhi

Risbud, N (1994) 'Property Market and Settlement Development: Case Study Rohini, Delhi', Research Report for HSMI, New Delhi

Singh, R (1999) 'Evaluation of Public Intervention in Resettlement of Squatter Settlements in Delhi', unpublished postgraduate thesis, Housing Department, School of Planning and Architecture, New Delhi

Wishwakarma, R K and Gupta, R (1994) *Organisational Effectiveness of Urban Basic Services Programme in Selected Slum Areas of Delhi*, Study sponsored by WHO, Centre for Urban Studies, Indian Institute of Public Administration, New Delhi

Chapter 4

Security of Tenure: Mumbai's Experience

Minar Pimple and Lysa John

Mumbai, previously known as Bombay, is expected to become the seventh largest urban conglomerate in the world, with a population estimated at well over 15 million. Studies undertaken by the government and social scientists estimate that between 55 and 60 per cent of Mumbai's population reside in slums, informal settlements characterized by structural and environmental degradation. It is estimated that approximately 40 per cent of slum households have an income below the poverty line.

The city of Mumbai is composed of a coastal stretch of 603 square kilometres. It is the prime destination of a steady number of migrants from rural areas and small towns in search of better economic prospects, or lured by the promise of an enhanced lifestyle. Population density in the city stands at a staggering figure of more than 17,000 people per square kilometre.

Geographically the city of Mumbai can be seen as divided into three sections, namely, the 'island city' (main city), the western suburbs and the eastern suburbs. These are also known as division I, division II and division III, respectively, for the administrative purposes of the local municipal corporation.

Division I represents the 'island city'. The earliest developed areas of the city fall in this division. It is a distinctive part of the city; a distinction based on characteristics of affluence, power and heritage. Land prices are at their highest here, for both commercial and residential purposes. Nariman Point holds the dubious distinction of being among the most expensive land holdings in Asia. Also situated here is the largest slum in Asia, Dharavi, which is a hub of industrial and entrepreneurial activity ranging from household enterprises like papad-making to centres for leather export. The Parel area is dotted with (now largely defunct) mills and workshops.

Commercial and residential developments mark division II. Certain pockets of this region are known for their affluent residential complexes. It also holds large slum pockets such as in Jogeshwari. The largest administrative ward of

Mumbai, K-east, is also situated here. On a surface level it appears to suffer less spatial saturation (as clearly visible in the island city). This accounts for increasing interest in this area by parties looking to invest with a residential or commercial purpose.

Division III is characterized by an industrial belt and a subsequent hazardous living environment. Areas such as Kurla and Chembur are infamous for their dangerous levels of air pollution. Both areas also house a range of factories and a vast slum population. Few areas are known for their hospitable residential localities catering to the middle classes.

The functions of planning and implementing land distribution and housing policies are divided among several city institutions, namely the Municipal Corporation of Greater Mumbai (MCGM), the Maharashtra Housing and Area Development Authority (MHADA), the Mumbai Metropolitan Region Development Authority (MMRDA), the collector and the judiciary.

Types of tenure arrangements in the city of Mumbai

Tenure in the city of Mumbai represents a direct relation between affordability and subsequent access to adequate and secure housing. Tenure relationships are historically of a tenancy and lease nature because a large proportion of the population cannot afford to actually own their place of residence. Approaches such as shareholding are being explored as alternative programmes for social housing.

There are three types of tenure arrangements, namely, tenancy, ownership and leasehold. The Bombay Rents, Hotel and Lodging House Rates Control Act of 1947 governs tenancy. It established a standard and reasonable charge to be paid by tenants and provided for controls against eviction of tenants by landowners. The provisions of the Act and its imminent amendments were factors that critically affected the security of tenure for the large numbers of tenants that reside in *chawls* (row tenements). These, by and large, are private holdings, with the exception of especially built *chawls* such as those developed by the Bombay Improvement Trust (BIT), known as BIT *chawls*.

The Maharashtra Rent Control Bill of 1993 is a proposed amendment to the Act. It is a state level adaptation of the model rent law, formulated as part of the national housing policy by the central government. It takes into account issues causing reduced investments in rental housing and the deterioration of existing rental housing stock. The proposed bill states that any person in occupation of premises on 1 February 1993 be recognized as a tenant of the landlord for the purposes of the Bombay Rent Act. Logically speaking the Act must apply only to instances of 'ownership', that is, to private property. However, by some deviation of application it also pertains to the 'illegal' landowner–tenant relationships in slums.

Another variance in the system of tenancy is the 'pugree system', which involves a mutually agreed upon, illegal transfer of tenancy rights. Thus, in an agreement reached between the landlord, the original tenant and the prospective tenant, the transfer of tenancy is arrived at by the transaction of a lump sum, payable in cash, of which one-third accrues to the landlord. Such a transfer allows for the 'new tenant' to gain superficial recognition as an original tenant and thus protection under the provisions of the Bombay Rent Act.

Ownership falls into two categories: freehold land ownership and shareholding. The Urban Land (Ceiling and Regulation) Act (ULCRA) of 1976, which stated that no individual (subject to certain exemptions in the cases of charitable and religious trusts and places of public interest, etc) could own more than 500 square metres of vacant land, influenced freehold land ownership (individual owners of land). Shareholding, the collective ownership of land through the formation of cooperative housing societies, is prevalent in the formal housing sector. It is introduced into schemes for subsidized housing to enable slum dwellers to benefit from improved housing facilities and to promote their participation in the process.

The third type of tenure is leasehold. Leases are provided on land applicable to an individual or on a collective basis. The lessee pays a lease rent. Collective leaseholds are applied to schemes for housing the poor, such as the slum rehabilitation scheme. The eligibility of these communities of slum dwellers is determined by the application of certain criteria. If the criteria are met they become entitled to a lease on the land to which they are rehabilitated. The lease rent was thus paid to the concerned authority.

Assessment of the Tenure Status of the Urban Poor

Housing provided by the formal sector in the cities of India is generally too expensive for the urban poor and the number of units built is insufficient. The urban poor in Mumbai tackle their housing needs by squatting on vacant land. Such settlements may, over the course of time, be recognized as legal and provided with some security of tenure. Slums currently listed in the census of 1976, or registered in the elections lists of 1985, are entitled to such recognition and benefits in terms of the provision of basic amenities. Early settlers may also rent part of their houses to newer arrivals in the city. Some may construct *chawls* that are leased on a unit basis. Such an arrangement can lead to a complex relationship between the *chawl* owner and the local authority.

The second option for the urban poor is the social housing schemes of the state. These are mainly directed towards the regulation of the land market, the equitable distribution of land, housing and services for the poor and slum and pavement dwellers. However, the coverage of the schemes is extremely limited in comparison with the magnitude of the need for adequate and affordable housing in the city. Moreover, procedures for eligibility seem to be based on the intent to exclude, rather than on the principle of housing as a universal right. These schemes have an uncertain future as they are subject to the fluctuations of power at the state legislature.

The policies for decongesting land and/or making it available for social housing have not been able to deal with the complexities of implementation –

Table 4.1 *Housing Arrangements of the Urban Poor*

1 Slums	Total land area under slums is 2525 ha or 26,260,000 m^2. Nearly 55% of the city's population lives on 6% of the land area of the city (Urban Design Research Institute, 1999) The Maharashtra Slum Areas Act of 1971 governs them. This provides some protection against eviction, as well as for basic amenities for slums declared official under the slum censuses of 1976, 1980 and 1985	The Act is the basis for schemes for slum improvement, upgrading and rehabilitation, outlined in the section of this chapter on the assessment of the tenure status of the urban poor The first slum census of 1976 also involved the issuing of photopasses to slum dwellers in the city; an identity card that shielded them from impromptu evictions and entitled them to the provision of basic services through the legal status that, for the first time, was bestowed on them
2 Pavement dwellers	Entire communities (known as *zhopadpattis*) housed in shanties alongside the footpaths, in open spaces, under bridges, etc Includes other destitute categories such as street children. Characterized by minimal access to amenities and social facilities. Minimal security of tenure One estimate puts the number of such homeless in the city at 980,000 (Nangi, 1995)	Such settlements are largely found on public lands or else on vacant spaces. In 1985 a supreme court judgement upheld the right of civic authorities to demolish unauthorized dwellings, leading to the continuation of large-scale evictions. Major evictions at the city level have been executed since 1981. In 1996, the Bombay Municipal Corporation (BMC) launched a massive drive wherein 27,058 households were rendered homeless
3 Settlements on vacant land	Regulated by the then Maharashtra Vacant Lands Acts of 1975 and 1980	The Act prohibited unauthorized occupation on vacant lands and empowered local authorities to evict illegal occupants. At the same time it negated any tenant–owner relationship on such lands. The later amendment provided further protection to dwellers from eviction and harassment. It also directed for the provision of basic amenities to slums on such lands at a fixed compensation rate
4 Displaced populations	Displacement is a prevalent phenomenon in the city. Slum and pavement dwellers have been perpetual victims of large-scale evictions; an apparent duality on the part of the government that protects the provision of shelter to the poor on the one hand and carries out aggressive demolitions on the other 10–20% of the slum dwellers whose communities are targeted	The combination of decreasing need for industrial labour and the increasing pressure on land for development is expected to cause massive displacement of the poor One group of people under the greatest threat of displacement in Mumbai city are workers who are presently employed in 'sick' or 'obsolete' industries such as the textile industry The Mumbai Urban Transport Project (MUTP), as conceived under the current

for redevelopment under the market incentive housing schemes are also anticipated to be displaced due to narrow criteria for eligibility Tenants in the island city also face displacement given the possibility that landowners will want to convert poor paying residential units (under the current rent norms) into commercial units for better returns	development plan, is estimated to displace 30,000 households. However, the 'Sukhtankar Committee' formulated under the MUTP has undertaken an attempt to look into the systems involved in relocating the displaced The recommendations of the committee involve the provision of a 30 year lease on the site of resettlement to the displaced communities and specifications to the effect that relocation must take place within a specified (proximate) radius from the displaced site Related issues such as provision for the transfer of home/community-based economic activity have also been considered in an attempt to make the package for rehabilitation more realistic and integrated Tenants in the displaced communities have also been allowed tenure rights in the resettlement sites. The mechanism of joint ownership of property (both men and women are recognized as owners) has also been incorporated

from the earliest 'development plan of greater Bombay' in 1964, to the most recent slum redevelopment scheme (SRD). While the agendas of the development plan and the ULCRA of 1976 have been contradicted in practice, schemes for slum improvement, upgrading and rehabilitation (having progressed in that order) have failed to meet their objectives. Ironically this has been attributed to the absence of community ownership and involvement in implementation, as characterized by the approach of the 1970s, and to the inability of the benefiting communities to participate cohesively – identified as a prerequisite for rehabilitation in the 1980s and 1990s.

The recommendations of the Sukhtankar Committee with regard to the displacement of communities by the MUTP are by far the most progressive for rehabilitation in the city. Its significance is in tackling the tenure problems of the larger groups of urban poor.

Interventions for security of tenure

Interventions in the area of tenure for the urban poor can be traced to initial attempts to help slum dwellers to improve their living environment, for example, by providing information and technical assistance to access credit or the mandatory clearances.

It was only after the period of 'emergency' (post-1975) and the subsequent curtailment of fundamental rights that an era of activism in the area of housing emerged. This period was marked by a change in ideology and functioning of several non-governmental organizations (NGOs). They moved from service-delivery and supportive roles to development-oriented roles, often involving issue-based advocacy.

The large-scale demolition of urban poor settlements in 1981 heightened the activism around issues of tenure. The plight of the slum dwellers was represented at the level of the Supreme Court (the Apex Court). The failure of the urban authorities to provide shelter for the poor – a fundamental right – was highlighted. The resulting judgement ordered the evictions to stop. This was a significant victory for the urban poor and the champions of their cause.

At present voluntary sector interventions on issues of tenure rights concern:

- processes aimed towards influencing amendments in existing policies or recommendations for the formulation of alternative policies, in a manner that is socially just and protects the interests of the poor;
- agitations against demolition activity and the protection of tenure rights of the urban poor; and
- building capacities within communities to negotiate projects of resettlement with government authorities.

The prominent NGOs working on the issues of housing are the Nivarra Hakka Suraksha Samiti (Society for the Protection of Basic Rights), the Society for the Promotion of Area Centres (SPARC) and Youth for Unity and Voluntary Action (YUVA). A city level network of organizations working on housing issues called the Mumbai Nagrik Vikas Manch (Citizen's Development Forum of Mumbai) also exists. There are instances of community-based organizations (CBOs) affiliated to political parties. This involvement in local leadership has facilitated access to government schemes, such as the SRD, opposition to demolitions and access to basic amenities.

Case Studies of Intervention Around Issues of Tenure

The following are instances of intervention as undertaken by YUVA. However, they are not exclusive in intent and reflect the nature of activities and responses to the issues of the urban poor as developed by the voluntary sector of the city.

The issues of tenants in slum communities: Janata squatters' colony, Jogeshwari

The mid-1950s in Mumbai were characterized by increased pressure on land, combined with a lack of affordable housing. These were the main reasons for the relocation of 1957 families from the prime areas of South and Central Mumbai to open patches of land measuring 15 feet x 20 feet within the Janata squatters' colony. The resettlement was facilitated by various schemes for development, including the Slum Clearance Scheme of 1956.

The new settlers were granted a legally binding title deed by the BMC, known as a vacant land tenancy (VLT). These VLT rights were conferred on four conditions, namely:

- Allottees were not permitted to expand their occupational areas beyond 15 feet x 20 feet.
- Allottees were not permitted to create subtenancies. However, the subtenants would be tolerated if the VLT rights were transferred, in their favour, to the respective area. Therefore subtenants would also become direct tenants of the BMC.
- Allottees were not permitted to sell their structures or transfer tenancy rights without official permission from the BMC and compliance with the due process of the law.
- Allottees had to pay a modest monthly rent to the municipal authority.

The conditions laid out in the VLT were violated by a majority of settlers. The land in the colony was quickly encroached upon by both old and new residents for purposes of tenancy or own family use. At this time, rent was collected from the subtenants and the receipts of transactions kept by the VLT holders. The area was left to grow unchecked until the introduction of the Maharashtra Vacant Lands Act (MVLA) in 1975. The Act, besides empowering the government to evict anyone on vacant lands, also nullified the relationship between tenant and owner by emphasizing the illegality of claims to ownership. It introduced a system of collection of compensation from all residents on these lands and issued photopasses to them. As a result the *chawl* tenants stopped paying rent to the VLT holders/*chawl* owners. From this time on the relationship between the *chawl* owner and tenant, which had previously been non-confrontational, rapidly deteriorated.

The MVLA was finally abolished in 1985 after a series of events and one amendment. This move was intended to renew and acknowledge the vacant land dwellers' legal status and rights. In some instances their ownership was officially recognized. They were allowed to collect rent and, in principle, were protected from evictions. This resulted in the reintroduction of the previous relationship between the VLT landholder/*chawl* owner and tenant. The *chawl* owners demanded that the tenant paid rent as well as rent arrears for the previous ten years, irrespective of the compensation they had been paid by the BMC. Different *chawl* owners filed cases in the small causes court to recover lost rent. Tenants were forced to legally defend themselves on an individual basis.

The verdicts on these cases were usually in favour of the *chawl* owner and the tenants had to pay rent to the owner on condition that basic facilities were provided. This verdict was in contradiction of the specifications of the Rent Act that disallowed the collection of rent for government land. Tenants who did not comply were evicted. The owners also resorted to illegal methods in a bid to recover lost rent; for example, lockouts, arson, assaults and threats, disconnection of water and electricity supplies, denial of permission to repair huts and so on. Tenants stopped improvements to their dwellings for fear of imminent eviction.

In 1990 there was collaboration between the BMC, YUVA and the Jogeshwari Rahiwasi Sangh (Jogeshwari Residents' Organization) (JRS). It was started as a response to the above-mentioned and related situations. The collaboration sought to address the legal contradictions inherent in the situation. It also aimed to work towards acquiring legal respite for tenants and to advocate their access to basic amenities.

An analysis of all policy documents available from the BMC was undertaken which clearly indicated that the VLT holders had violated the conditions of the VLT agreements of 1956. A character survey was undertaken. This required plots to be measured and then compared with the original allotted sites of the areas in order to assess the extent of violation of the VLT rules. The JRS and YUVA also provided the tenants with training in legal literacy so that they could effectively interact with their lawyers and participate in the cases.

The BMC expressed its commitment to resolve the dispute and bring clarity to the inherent contradictions in the situation. It also undertook a similar character survey, along with the *chawl* owners, to assess the number of plots of land and discrepancies in relation to the original VLT allotment area. Thus, it provided the opportunity for both parties, the *chawl* owners and the tenants, to be fairly represented in the legal proceedings. It also readily provided material from the government archives for review by the JRS and YUVA. This contributed significantly to the progress of the case. This is an excellent example of how access to information, currently regarded as a privilege rather than a public right, can have a positive outcome.

The Deputy Municipal Commissioner (DMC) of zone III heard the case. He had been empowered to exercise quasi-judicial powers by the High Court in 1990. He adjudged that the VLT holders had violated their original VLT conditions and were therefore illegal encroachers on the land without permission to rent out property. This in itself was enough to revoke their VLT status. Moreover, since slum improvement schemes were applicable to the colony, the status of all occupants, whether owner or tenant, was declared the same. The BMC's law department was also asked to intervene in the ongoing legal disputes between tenants and *chawl* owners in the small causes court. The respondents found this part of the judgement the most significant.

However, on the downside, the premature transfer of the hearing resulted in insufficient institutionalization of the status accorded to the tenants in the case, thus preventing its expanded application to other tenants facing similar predicaments.

Advocacy leading to resettlement: the case of Bhabrekar Nagar

In June 1977, at the height of the monsoon, 12,842 families were rendered homeless due to massive demolitions in Bhabrekar Nagar, Mumbai. A comprehensive advocacy strategy was devised by YUVA to explore the issue. This included a survey of the community, the gathering of information and proof of residency. This helped in building rapport with the community and it served as a crucial tool in the subsequent negotiations with the government.

A public hearing, attended by over 1000 people, revealed that some people had died and a woman had given birth to her child on the road. A fact-finding mission conducted by the Habitat International Coalition (HIC) presented this information to the chief minister, who committed himself to investigate the partial illegality and social injustice of the demolition. As a result of such lobbying the state government allotted a plot of land for resettlement and sanctioned an amount of Rs9,000,000 (approximately US$225,000) for basic amenities. This was the first grant given for an urban resettlement project for people affected by demolitions.

Public interest litigation in the matter of the implementation of the SRD at Jogeshwari, Mumbai

In 1995 a petition against the implementation of the SRD in Potters' colony (known locally as Shivaji Nagar and Hari Nagar) was filed for the following reasons:

Provoking communal tension

The scheme was said to have selectively sought to evict slum dwellers belonging to the Muslim community on the said land, although they were in every way entitled to participate in the scheme. This section of the slum dwellers was not included in the certified lists of those eligible for the scheme made by the BMC, nor did they sign the documents necessary for the builder and promoters to obtain the consent of the slum dwellers.

Violation of statutory provision of the scheme

The statutory provisions of the scheme mandated that every eligible slum dweller should be resettled on the land to be redeveloped. This was violated. The scheme also deprived the state of prime land and gave the benefit of the land and excess floor space index (FSI) to the builder.

The land in question was originally privately owned but had been taken over by the BMC for the purpose of resettling potters. Hence it became known as the Potters' colony. Over the years members of all castes and religions settled on the land. In accordance with government guidelines issued from time to time, they were given photopasses and obtained other civic amenities. From at least 1985 the residents have paid municipal taxes and have been included in the census and listed on the electoral rolls.

The petition was considered a matter of public interest because it was filed on behalf of 7500 residents of the said land. The violation of statutory procedures in the implementation of the scheme had bigger implications for the large population of slum dwellers. These slum dwellers stood to either benefit from or be deprived of such schemes for mass housing, depending on whether the petition challenged the emphasis of the builders' profits or, alternatively, placed the people at the centre of such schemes.

The main demands of the petition filed jointly by the hari nagar shivaji nagar jhopadpatti punarvikas samyukt samiti (slum redevelopment joint committee) and YUVA were:

- Protection of the right to permanent housing of slum dwellers who do not consent to the scheme.
- Prevention of the arbitrary exclusion of any eligible slum dweller. Certification of voluntary exclusion from the scheme in case any slum dweller does not want to be included.
- Complete disclosure of all details of the scheme. The slum dwellers' cooperative societies must also be given the power to supervise, inspect and certify the construction of tenements.
- Ensure that the open market tenements are sold at the prices stipulated in the scheme by inviting applications from the public and drawing lots to decide allotment.

As a result, the courts issued a directive to the government parties to institute an inquiry into the matter. The inquiry reinforced the fact of insufficient consent. The coverage of the scheme was thus restricted to three *chawls* within the larger community.

The petition highlighted the cross-section dynamics in the implementation of a seemingly comprehensive and democratic scheme. It questioned the premise that political will alone will suffice for effective implementation. Such will, according to some interventionists, is more likely to be influenced by the political and economic spin-offs for the implementing parties. These in turn are manifested in the relationships between community groups, thus negatively influencing the process of implementation of the scheme, as well as the protection of the security of tenure of the larger community.

REFERENCES AND BIBLIOGRAPHY

Charles, L (1993) *Securing a Foothold: A Case Study of the Tenants of Janata Colony Jogeshwari (E), Bombay*, paper presented at a workshop on NGO–Local Government Collaboration in Management of Local Development for South Asian Countries, Dhakka, Bangladesh, 29–30 December 1993

Eviction Watch–ACHR (Asian Coalition for Housing Rights) and YUVA (Youth for Unity and Voluntary Action) (1996) *Footnotes of a City: Housing Situation and Policy Alternatives for Pavement Dwellers*, YUVA, Mumbai

Eviction Watch–ACHR and YUVA (1997) *Inquest of Bhabrekar Nagar: A Report of the Fact Finding Mission constituted by the Habitat International Coalition to Enquire into the Demolitions at Bhabrekar Nagar, Kandivali, Mumbai,* June

Gehlen, N and Stenzel, W (1998) *Planning for the Urban Poor: A Case Study on Housing in Bombay,* project of the Asia, South America and Africa Programme of the Carl-Duisberg-Gesellschaft, Federal States of Germany and Ministry for Economic Cooperation, Bonn

Kalpana, S (1994) *Waiting for Water: The Experience of Poor Communities in Bombay: a Case Study,* Habitat International Coalition, Mexico

Meerburg, E S (1995) 'Land for Housing Bombay's Poor: Government Interventions in the Land and Housing Market to Support the Urban Poor', thesis, Faculty of Geodetic Engineering, Delft University of Technology

Nangi, P (1995) 'Urban Imbalance', *The Review*, Vol III, No 89, August–September, Forum for People-Oriented Water Utilisation (POWU), Mumbai

Urban Design Research Institute and Mumbai Nagrik Vikas Manch, 'Recommendations towards developing an alternative set of policies for the city's development', *FSI and Mumbai's Development*, Urban Design Research Institute, Mumbai

Vandana, D (1995) *Overview of NGOs in Bombay,* Institute of Development Studies, Brighton

YUVA (Youth for Unity and Voluntary Action) 'Travesty of Justice: An Appeal for Solidarity with Slum and Pavement Dwellers', Internal Document, YUVA

YUVA (Youth for Unity and Voluntary Action) (1999) *Our Home is a Slum: An Exploration of a Community and Local Government Collaboration in a Tenant's Struggle to Establish Legal Residency in Janata Squatters' Colony*, United Nations Research Institute for Social Development (UNRISD), UNU, Mumbai

YUVA (Youth for Unity and Voluntary Action) (1998) *Our Home is a Slum: An Exploration of a Collaborative Initiative between Local Government, a Voluntary Organisation and People's Organisation,* Janata Squatters' Colony, Mumbai, April 1998

Chapter 5

Security of Tenure of Irregular Settlements in Visakhapatnam

Banashree Banerjee

INFORMAL SETTLEMENTS IN A RAPIDLY GROWING CITY

Visakhapatnam or Vizag is situated on the east coast of India, midway between Calcutta and Madras. It is the second largest city in the state of Andhra Pradesh and it ranks twenty-first in size and first in growth rate in the country. Between 1971 and 1991 its population trebled from 360,000 to over 1 million. This rapid increase reflects the city's growing importance as a major port and industrial centre, as well as the poverty of the surrounding rural areas from which many migrate. The city lies between the hills and the sea, a situation that has created a constraint on land. Population density is high, with 30,000 people per square kilometre in most of the city. The premium on land and the building boom, dominated by private builders, is rapidly turning the city into a mass of high rise blocks into which businesses and upper and middle class families are moving. Even so, housing for employees of the railway, port authorities, government departments and public sector industries occupies a major portion of residential land in Vizag.

Although approximately 60 per cent of land within the municipal limits of Vizag is owned by public organizations, only a very small share of this is available for public use. The largest holder of government land (40 per cent) is Visakhapatnam Port Trust (VPT), followed by the government of Andhra Pradesh (GOAP) revenue department and Indian Railways. Railway and VPT land is consolidated, while revenue land is scattered in small pieces and is mostly in the form of road and drain reservations, beaches and hill slopes. These areas are classified as 'objectionable' from the perspective of habitation, but they have provided the only viable housing option for a large number of poor families.

So, despite industrial prosperity, or attracted by it, 240,000 people or 52,000 households live in 251 officially designated slums, with an average annual

Table 5.1 *Distribution of Slums by Original Land Ownership in MCV area, Vizag*

Landowner	*Slums*		*Remarks*
	Number	*%*	
State govt/MCV	132	53	
State govt departments and other institutions	7	2.5	VUDA, Dairy, Health Dept, Andhra Pradesh Housing Board
Central govt	19	7.5	VPT, railways, defence
Religious trusts	10	4	Church, wakf, endowment
Private	60	24	8 have some govt land; in 18 inner-city slums plots are owned by residents
Village	23	9	Plots are mostly owned
Total number of slums	251	100	

Source: MCV records, 1998

household income of Rs10,500 (US$244). They occupy road and railway margins, hill slopes, sides of large city drains (gedda), beaches and other left-over land in the city, and crowd into the old city.

The city is also growing outward into the surrounding rural areas and joining up with smaller towns in the Visakhapatnam Metropolitan Region (VMR). The fringes are rapidly developing with industries, shrimp farms and housing in the form of Visakhapatnam Urban Development Authority (VUDA) townships and private residential layouts, both formal and informal.

City planning and development: who does what?

The Municipal Corporation of Visakhapatnam (MCV) is responsible for all municipal functions within its 111.6 square kilometre area. The MCV is headed by the mayor and has a council of elected ward councillors. The commissioner looks after the administration and technical departments of the MCV. Poverty alleviation and slum improvement are among the functions of the MCV. On the other hand, the Municipal Corporation Act also authorizes the MCV to demolish irregular structures. Administration of building regulations is a municipal function, while zoning and subdivision regulations are the responsibility of VUDA, which has the responsibility for development planning and control in the VMR.

In 1989 GOAP approved the master plan for VMR up to 2000. This covers an area of 1721 square kilometres and includes three other towns besides Vizag and 287 villages. Apart from its planning and regulatory function, VUDA acquires, develops and allocates land for various purposes. With regard to informal settlements, VUDA has prepared and approved layouts for re-blocking slum land in order to implement tenure regularization and housing programmes. This responsibility has been shifted to the MCV. VUDA also takes punitive measures against unauthorized layouts, as well as regularizing them.

Public housing is the responsibility of the Andhra Pradesh State Housing Board (APSHB) as far as all but the poor are concerned. The Andhra Pradesh State Housing Corporation Ltd (APSHCL) takes up housing for the urban and rural poor. Both the organizations function through their divisional offices and depend on funding from GOAP and housing finance institutions. The APSHCL has executed a number of house construction and shelter upgrading projects in Vizag over the past two decades. It has confined its operations in Vizag almost entirely to slum and relocation areas.

The revenue department, though not associated with urban development, is important because of its responsibility for administering state lands and all property transactions. The department regularizes tenure of poor settlers on government land by granting *pattas*. It is also responsible for the eviction of unauthorized people occupying government land.

Informal settlements: typology

The two main types of informal settlements in Vizag are slums and unauthorized layouts. The information on slums and measures to improve them is well documented, but little is known about unauthorized layouts. It is for this reason that this chapter concentrates on slums and only makes passing references to unauthorized layouts.

In Vizag (as in most Indian cities) the term 'slum' is very loosely used. It includes a variety of substandard areas such as squatter settlements, inner city slums, villages, illegal subdivisions and relocation areas. However, they do conform to the definition of a slum given in the 1956 Andhra Pradesh Slum Improvement (Acquisition of Land) Act:

> *Where the Government is satisfied that any area is, or may be, a source of danger to public health, safety or convenience of its neighbourhood by reason of the area being low-lying, unsanitary, squalid or otherwise, they may, by notification in the Andhra Gazette declare such an area to be a slum area* (section 3(1)).

Even though informal settlements are not differentiated by nomenclature, there is a difference in the nature of interventions related to them. Over the years 95 per cent of these slums have received some form of secure tenure, legal or otherwise.

Slum improvement and tenure regularization: Two decades of experience

According to statistics, conditions in Vizag slums have undergone a remarkable change for the better: shelter conditions have improved; tenure is less precarious; people have better access to civic infrastructure; literacy rates have gone up; access to preventive health care has improved; and more than half of those who underwent vocational training are employed. These improvements can be attributed to a number of programmes implemented in Vizag since 1979.

Different researchers and evaluation teams have recorded other benefits from the interventions. Infrastructure and shelter improvement and tenure security have increased land values and rents of slum properties. Community perceived benefits include: a nice environment leading to higher social status; regularization leading to freedom from exploitation; and water supply saving time and reducing drudgery (Development Action Group, 1997). The benefits of infrastructure improvement are sustained in a number of slums through community participation in the operation and maintenance of services such as boreholes, streets and drains. Women are making local decisions and approaching other institutions for their requirements (Subramanyam, 1999).

An integrated and intersectoral approach to slum improvement

The remarkable feature of slum improvements in Vizag is the unified and integrated approach that has consistently been followed. The prime responsibility for slum improvement lies with the MCV. In 1979 the urban community development (UCD) programme was started and continued up to 1986.

During its seven years of implementation the programme instituted the culture of comprehensive development of slum communities by converging resources and programmes and coordinating between different organizations. Through this approach it was possible to include and bring together the following: the housing programmes of the APSHCL; tenure regularization programmes of GOAP implemented by the revenue department; infrastructure improvement by the MCV (as part of the 'environmental improvement of urban slums' (EIUS)); and other health, education and training programmes being implemented by different government and non-governmental organizations (NGOs). The community development approach facilitated the formation of neighbourhood committees (NHCs) and gave them space to participate in their own development (Asthana, 1994).

The Department for International Development of the United Kingdom (DFID) funded the Visakhapatnam slum improvement project (VSIP) from 1988 to 1995, instilling the principles of the UCD on a city-wide scale. It included large grants for infrastructure improvement, for which resources were a major problem during the UCD phase. The budget for social, economic and health activities was also substantially higher. This made the operations of the urban community development department (UCDD) more self-sufficient, but with institutional expansion and diversification. Even then, the need for coordinating inputs from other organizations remained a major task, as housing and tenure regularization were not included in the VSIP.

The participatory approach was further strengthened with the DFID-supported Chinagadili Habitat improvement scheme (CHIS) from 1993 to 1998, which was implemented in a large relocation area. The NHCs participated in preparing action plans for the improvement of their own area. Cross-cutting monitoring committees were set up and partnerships were established with local NGOs to implement the women's savings and credit programmes and education programmes. Tenure and housing programmes started even before CHIS and continued parallel to it.

The current Government of India programmes being implemented are the swarna jayanti shahari rojgar yojna (SJSRY) and the national slum development programme (NSDP), which also advocate a participatory and integrated approach.

Mechanisms for institutional development and coordination

The UCDD has emerged as a nodal institution in relation to all interventions in slums. More importantly, in terms of Indian institutional politics, other institutions accept its role. At a strategic level, the project director maintains contact with other institutions as a member of the decision-making committees for *pattas* and housing, and also represents the MCV in all inter-institutional matters relating to slums. From time to time necessary powers are delegated to the project director for this purpose. However, he or she has limited financial powers, and delays in implementation often result because of the length of time taken to get administrative and financial approval.

The UCDD's dealings with the public, government and NGOs, as well as a large budget, have necessitated transparency of operations and meticulous accounting. Remaining non-controversial, although not always possible, has been one of the keys to its political survival.

At field level, community organizers have provided the link between community groups and public institutions. They inform community groups of the various programmes and advise and inform individual families. The UCD field staff, assisted by grass-roots workers, have carried out surveys required to identify beneficiaries for various programmes and involved the NHCs in the process. These activities are particularly valued by organizations like APSHCL and the revenue department, which depend on UCD for field coordination.

Creating and keeping communication channels open has been central to the process of coordination. Gaps have invariably developed whenever, for whatever reason, there has been a disruption. Over the years an unwritten code of practice has developed to everyone's advantage, except perhaps the local leaders who have often tried to disrupt the process. In fact in 1992, because of corrupt practices of beneficiary selection by councillors and local leaders, the shelter upgrading programme was transferred from the MCV to APSHCL.

The training and capacity building component of the UCD, VSIP and CHIS have provided opportunities for improving the skills and awareness of staff. Staff were trained at different institutions in India and abroad. They also went on field visits. This exposure gave them confidence and knowledge about their work. Capacity building of community leaders through training and exposure visits formed vital components of the UCD, VSIP and other programmes.

Learning from Hyderabad and moving forward

In many ways the smooth operation of slum improvement efforts in Vizag was due to the lessons learnt from the Hyderabad UCD programme and procedures developed for it. Most of the good practices of Hyderabad were replicated. The transfer of experience was possible at all levels because of field visits, training programmes and documentation. But even more important was the formalization of the UCD approach in clear administrative guidelines and procedures for all government institutions and funding mechanisms. The close relationship between tenure regularization, shelter construction and infrastructure development, which was possible right from the inception of the slum improvement programmes, can be seen as a result of the acceptance of the convergence approach in Andhra Pradesh.

SECURITY OF TENURE IN SLUMS: THE DIFFERENT PATHS

The latest records show that as many as 57.5 per cent of the slum population of Vizag have been granted some form of *patta*. Moreover only 5.5 per cent of people are living in conditions of insecure tenure. Those in-between may not have legal claims over the land they occupy, but their tenure position is safeguarded by a variety of positive or negative factors (see Table 5.2). The positive factors can be considered as those that result out of some form of official intervention or investment (such as infrastructure improvement, assurance of relocation or identity cards). Negative factors are those that make the land unfit for any other use and, by default, tenure is secure (examples are poor site conditions, property disputes and litigation).

Table 5.2 *Number of Persons Benefiting from Improvement Measures in Vizag Slums*

Nature of intervention	*Persons*		*Remarks*
	Number	*%*	
Housing loan, *patta*, infrastructure improvement	88,062	37	Housing in progress for 6334 of these
Shelter upgrading, possession certificates, infrastructure improvement	6243	2.6	
Title deeds and infrastructure	1650	0.7	
House site *patta*, infrastructure improvement	27,907	11.8	Issue of house site *pattas* going on for 2663 of these
Infrastructure improvement	39,877	16.8	
Patta	12,608	5.4	Of these, *pattas* approved for 1230, to be issued
Identity card	205	0.1	
No intervention but other indications of secure tenure	47,832	20.1	Own plot, proposed for relocation, disputed property, litigation
No intervention, no tenure security or official infrastructure improvement	13,053	5.5	On private, railway and defence land, endowment land
Total slum population in Vizag	237,437	100	

Source: Compiled from MCV and APSHCL records 1999

Different kinds of tenure documents are issued depending on the specific situation and are upgraded as and when necessary. House site *pattas* are granted by the GOAP revenue department in accordance with Government Order No 508 of 1995, applicable to all urban areas of the state. All those who have been living in a house on government land for more than five years are eligible to apply for *patta*. Applications are scrutinized by a committee consisting of the revenue department, MCV and VUDA and *pattas* are given to recommended cases. In some cases provisional *pattas* are issued by the municipal commissioner, to be replaced by regular *pattas* later. Identity cards may be issued by the MCV or revenue department to establish occupation of land on a particular date for issue of *pattas*. When *pattas* are mortgaged against housing loans, possession certificates are given to guarantee rights over the land until the loan is paid off. They also record loan recovery. Title deeds have been issued only in villages that have come into city limits and are notified as slums.

Every *patta* document mentions conditions under which tenure can be retained. These relate to the underlying objectives of the government in taking up the programme. Granting of legal land rights to the poor at a nominal cost, or no cost at all, is seen as a welfare measure intended to allow security of tenure and not to turn the plot into marketable property. To fulfil this purpose *pattas* are not transferable and are for residential use only. The Andhra Pradesh Assigned Lands (Prohibition of Transfer) Act of 1977 makes it mandatory for the registrar of properties to ensure that no transfers of *pattas* are registered. Another method is to issue *pattas* in the name of women. Experience has shown that women are more interested in secure tenure than financial gains from selling plots. In spite of these conditions, there is still a market in slum lands.

Land ownership and tenure conferment

It has been relatively easy to provide *pattas* for state government and municipal land because the procedure is simple and third party interests in such land are limited. Slums on 'objectionable' land, however, pose difficulties. These are considered unfit for building because they are environmentally precarious. The earlier practice was to relocate residents to safer places by granting tenure in alternative locations. There are a number of such small relocation pockets scattered in the city. But now the availability of suitable land is highly constrained and the current policy is to issue *pattas* on objectionable land such as steep hill slopes, *geddas* and beaches. The only cases where relocation from government land is now taking place is where the land is required for essential infrastructure, or where the reconstituted layout for undertaking the housing programme cannot accommodate all families. This is being avoided by building two- to four-storey housing blocks, instead of the earlier single-storey row houses.

Success on private and trust land is more limited. The land must first be acquired under the Slum Act, a procedure that takes many years. The procedure started in five slums in 1984 is not even halfway through. Moreover, cases run into litigation over boundary and ownership disputes, a condition that is endemic to all urban areas in the country because of poor land records. Only three of these cases could be resolved by negotiating land sharing arrangements between the community and the landowner. Most slums on private land are in the old city with its easy access to employment, and for this reason slum dwellers are reluctant to move out. In any case legal proceedings lend immunity from eviction and services can be easily procured from informal sources inside the city.

Tenure regularization on central government land is even more difficult. After prolonged negotiations the VPT agreed to transfer slum lands to the MCV. Slums on railway and defence lands are still without physical infrastructure improvements. Social and health programmes have been extended to them and some services have been provided informally by ward councillors. Negotiations are still going on for the transfer of land.

Linking tenure regularization with housing and infrastructure

The link between tenure regularization, housing and infrastructure is one of the success stories of slum improvements in Vizag. The three components have had a substantial impact on improving living conditions, especially where plots have been re-blocked. Overall improvement and the promise of a 'nice place to live' have induced families to make substantial investments in housing, over and above the loan from the Housing and Urban Development Corporation (HUDCO). In many cases the extra investment has been used to create space for renting or for home-based economic activities, thereby improving family income (Rao and Ramachandrudu, 1992), but according to *patta* conditions, these economic activities are not permitted.

The process of gentrification through the sale of plots and houses to higher income families is reported to be quite substantial, particularly in well located hill areas along the national highway. The increase in property value rates in these areas has surpassed those in the city (Abelson, 1996). Some of these sales are 'distress sales' by families that have fallen into the debt trap because of their initial enthusiasm to build a nice house. There are cases of borrowing amounts three to four times the HUDCO loan from private moneylenders at high rates of interest. When repayment becomes impossible, sale of the plot and house, sometimes unfinished, is the only solution (Asthana, 1994). Of course, others follow the market to make a profit. This illegal sale of plots has given rise to its own set of middlemen promising high profits or protection from the law. Local leaders are in the forefront of the business. There are some leaders who have given up permanent jobs to pursue this activity (Asthana, 1994).

Relocation is another fallout of the housing–tenure link. As already mentioned, it is not always possible to accommodate all families on the reconstituted layout. Through a process of community consultation, families are asked to volunteer for relocation to another site. In some cases the decision is not voluntary but is forced upon the weaker members of society, the poorest and the low caste people (Asthana, 1994).

In the past there have been delays in the mobilization of funds for housing programmes. This has led to late starts or stopping of construction work midway, causing a lot of uncertainty and worsening living conditions because the original houses have been demolished.

The impact of community development on tenure regularization

A number of advantages have emerged. The process of dialogue in the community, facilitated by the community organizer, resolves several issues such as whether to opt for a housing programme, and who is willing to move out if the layout cannot accommodate all families. The UCD staff and educated members of the slum community have been assisting others to fill in forms and complete other formalities with APSHCL. When the self-help housing programme was being implemented, there were several cases of community members managing and participating in the building process. Well-organized community groups are also undertaking the operation and maintenance of services (Subramanyam, 1999).

The turnover of plots is much lower because of the well-organized and active women-dominated NHCs. However, as soon as the housing programme is implemented, leadership is taken over by male members, who also often act as property brokers. They are known to charge 'fees' from community members for facilitating all sorts of activities, supposedly because of their closeness to UCD staff and political bosses (Asthana, 1994).

Case studies of tenure regularization in some slums

In situ tenure regularization on state government land

Jayaprakash Nagar, Ganesh Nagar and Lakshmi Nagar are all hillslope slums along the national highway. Jayaprakash Nagar consists of 103 families who were all accommodated in the self-help housing scheme layout in 1982. The community organization was started before even the UCD programme and has been very active in organizing community members to construct a community hall with a contribution from the people. The community is now associated with MCV in maintaining services. Plots have been sold but not on a significant scale. The Jayaprakash Nagar's NHC is considered a 'model' (Subramanyam, 1999).

In contrast, in Ganesh Nagar, with 507 plots, only 48 per cent of the original allottees were still living in the area by 1993. Twenty-six and a half per cent of houses were rented and 25.5 per cent were illegally sold. Families invested heavily in house construction, mostly by borrowing from private moneylenders. The original NHC was very active and carried out many activities; for example, it raised community contributions to pay for a preschool teacher. However, with the coming of *pattas* and the housing programme in 1982, another set of leaders emerged. They had contacts with a dominant political party and were managing the real estate business in the upgraded slum (Asthana, 1994).

Women formed the Lakshmi Nagar NHC as part of UCD. Even though internally strong, the group had difficulties in approaching the authorities in the initial years. *Pattas* and housing for the 104 families only came about in 1992 (Asthana, 1994). There has been no turnover of plots and houses according to UCDD sources. Investment in housing has been only marginally higher than the loan amount.

Relocation on government land

Chinagadili was developed as a site-and-services layout to resettle about 5000 families from slums on national highway margins, objectionable government land and families left over from in situ housing layouts. In two of the sectors, *pattas* were given at the time of relocation in 1986 and housing loans given between 1986 and 1993. However, soon after relocation, the land of the other three sectors came under litigation because a private individual claimed the land as his. *Pattas* and housing loans could not be given until the final decision of the land tribunal of GOAP came in 1998.

In the meantime infrastructure was provided as part of the CHIS and social and health programmes were undertaken. According to UCDD staff, the improvement of services, a good location and frequent bus service has made the area an attractive suburb, where even the middle class want to buy plots. Here too a network of local leaders dominates the property business. A survey carried out before *pattas* were given revealed a high turnover of plots. A policy decision was taken to grant *pattas* free of cost to original allottees and to charge the market price from newcomers. Local leaders, the UCDD and revenue department representatives formed a *patta* committee to monitor the *patta* process.

Negotiations for tenure regularization on private land

Nerella Koneru is a central city slum with 108 families, originally on private land. The declaration of the area as a slum under the Slum Act in 1979 was followed by acquisition procedures in 1980, when the landowner went to court against the proceedings. The court gave a stay order to maintain the status quo. In the meantime some other people laid a claim to the land and went to court. The original stay order was vacated but the second case continued. During this time the revenue authorities issued provisional *pattas* and a housing scheme was started with a bank loan. But it had to be stopped because the second court case opened. It was then that the MCV commissioner invited the landowner to negotiate a settlement to share the land. Negotiations started in 1982 and a deal was finalized in 1992. The landowner handed over 60 per cent of the land to the MCV for slum dwellers' housing. The MCV had to pay at the rate applicable for expropriation, which is below the market rate.

UNAUTHORIZED LAYOUTS AND THEIR REGULARIZATION

VUDA is responsible for the regularization of unauthorized layouts in accordance with a state government order issued in 1987 for the cities of Hyderabad, Visakhapatnam and Vijayawada and their surrounding rural areas.

The urban development authority is first expected to take penal action against illegal developers or plot holders. If that is not possible, the layout is to be regularized with suitable modifications and payment of charges for regularization and services provision. Building permission for individual plots can be granted only after regularization charges are collected from at least 50 per cent of the plots and land transfer is registered.

According to official sources, all unauthorized layouts are outside the MCV area. Two types of situations exist.

Prior to the formation of VUDA, the Gram Panchayats (village councils) were authorized to approve land subdivision for building purposes after obtaining technical approval from the director of town and country planning (DTP) of GOAP. This was a time-consuming process. In several cases the heads of the Gram Panchayats followed the irregular practice of issuing layout approvals without referring to the DTP. After the formation of VUDA there was an insistence that subdivision regulations are followed and proper approval obtained before infrastructure can be extended and building permission granted. Fifteen layouts on the outskirts of Vizag, covering about 40 hectares, have applied for regularization and are at various stages of the process in accordance with the GOAP order of 1987.

The layouts are in scattered locations and in some cases infrastructure extensions are seen as a problem. Moreover, the middle class plot holders are reluctant to pay for services before they are actually provided. There has been a high turnover of plots and a spiralling of prices owing to the outward expansion of the city.

The second situation is the illegal subdivision and sale of endowment land (land gifted to a Hindu temple and managed by trustees) of the Simhachalam temple, located in the city's periphery. About 200 hectares out of the 400 hectares leased for cultivation by the temple trust have been subdivided and sold illegally in various forms. There are very small plots (50 to 70 square metres) purchased mostly by industrial and port labour. Twelve layouts with larger plots have been purchased by registered cooperative societies. These layouts follow the standards of the subdivision regulations. There are also layouts with large plots, obviously for speculation or luxury housing. The issue of these subdivisions came to light when some members of cooperatives were denied building permission. In 1998 a committee was formed by the Andhra Pradesh legislative assembly to look into the matter. A decision has not yet been made.

The approach towards unauthorized layouts has been reactive, unlike the proactive approach to the regularization of slums. No systematic inventory of unauthorized layouts has been attempted by VUDA or any other agency. Even though the authorities claim that there are no unauthorized layouts in the MCV area, some settlements classified as slums are indeed illegal subdivisions with plots owned by the residents. These areas have been provided with services as part of VSIP.

REFERENCES AND BIBLIOGRAPHY

Abelson, P (1996) 'Evaluation of Slum Improvement: Case Study in Visakhapatnam', *Habitat International,* Vol 13, No 2, pp97–108

Asia Law House (1999) *Land Grabbing Laws in Andhra Pradesh,* Asia Law House, Hyderabad

Asthana, S (1994) 'Integrated Slum Improvement in Visakhapatnam, India', *Habitat International,* Vol 18, No 1, pp57–70 de Bok, T (1994) 'From Slum to Relocation Area', master's dissertation, University of Technology, Eindhoven

de Bok, T (1997) *Improvements for Whom? A Case Study in Chinagadili*, DFID-sponsored research report

Development Action Group (1997) *Impact Assessment Study of Slum Improvement Projects in India*, University of Birmingham

Government of Andhra Pradesh (1956) 'Andhra Pradesh (Acquisition of Land) Act', *Andhra Pradesh Gazette*, Part IV–B Extraordinary, 15 November, p1

Government of Andhra Pradesh (1975) *Andhra Pradesh Urban Areas (Development) Act*, for the Visakhapatnam Slum Improvement Project, Municipal Corporation of Visakhapatnam (MCV), 1987

MCV (Municipal Corporation of Visakhapatnam) (1991) *Chinagadili Habitat Improvement Scheme*, mimeo, MCV, Visakhapatnam

MCV (Municipal Corporation of Visakhapatnam) (1995) *Chinagadili Habitat Improvement Scheme–II*, mimeo, MCV, Visakhapatnam

Sarveswara, R B and Ramachandrudu, G (1992) *Evaluation Study of the Visakhapatnam Slum Improvement Project: Report on the Household Survey,* Institute of Planning and Development Studies, Visakhapatnam

Subramanyam, U (1999) 'Impact of Community Development Approach for Housing and Infrastructure in Slums: A Case Study of Visakhapatnam', *HSMI Research Report No 12,* New Delhi

VUDA (Visakhapatnam Urban Development Authority) (1990) *Master Plan – Visakhapatnam Metropolitan Region*, VUDA, Visakhapatnam

Part 2 Brazil

After the chapters in this section were written, three significant legal developments took place in Brazil: the Constitutional Amendment No 26 of 14 February 2000 incorporated the right to housing into the set of social rights originally listed in Article 60 of the 1988 Constitution; the National Congress eventually approved the national law on urban policy, entitled 'City Statute' (Federal Law No 10.257 of 10 July 2001), regulating Articles 182 and 183 of the Constitution, as well as creating new instruments for democratic city management and land tenure regularization; and on 4 September 2001 the President of Brazil signed the 'Provisional Measure' No 2.220 regulating the instrument of 'special concession of use for residential purposes' mentioned in Paragraphy 1 of Article 183 of the Constitution, and recognizing the rights of occupiers of public land.

Chapter 6

Providing Security of Land Tenure for the Urban Poor: The Brazilian Experience

Edesio Fernandes

Introduction

The combined processes of industrialization and urbanization in Brazil began in the 1930s, when less than 30 per cent of the population lived in cities, and over the years it has provoked dramatic changes in Brazil's socio-economic and territorial order, as well as having a serious impact on the environment. Today, after seven decades of intensive urbanization, about 80 per cent of the total population (estimated at 165 million people) live in cities, with over 40 per cent of the urban population living in metropolitan areas. Since the mid-1950s, most Brazilian wealth has been generated in cities. Even though there has been a significant decrease in both migration and urbanization rates over the past decade, the urban, and especially the metropolitan, population is still rapidly growing.

On the whole, the first stage of urban development in Brazil was largely uncontrolled, but did meet the requirements of the broader socio-economic development process. Prior to the promulgation of the 1988 federal constitution, there was no coherent planning and institutional framework in place at national level. To date there is no national planning law, and the limited federal legislation only refers to some aspects of the urban development process. A systematic approach to urban planning was only instituted at national level during the period of military rule in the 1970s, when urbanization peaked.

Urban planning from the early 1970s to the mid-1980s was characterized by the subordination of state agencies to private interests. Therefore, national urban development policies were in the interests of the landowners and large companies that developed urban infrastructure (Fernandes and Rolnik, 1998).

This was facilitated by the failure of the Brazilian state, prior to the constitution, to reform laissez-faire legal ideologies regarding the use and development of urban land.

The cities, which were increasingly the centres of growing industrial production, had to absorb the mass of migrants looking for job opportunities. However, despite the very high rates of urban growth, local government has done little to provide housing options and public services to the vast majority of urban dwellers. Before 1988, local planning was incipient and limited to the central areas where land use and development were regulated in basic legal terms. From the early 1960s until the late 1980s, local urban policies were also hostage to the strategies of large-scale urban developers and promoters.

Access to land and housing has been led mostly by the private sector, according to formal and informal processes. As a result, very peripheral and irregular land subdivisions (*loteamentos*) with self-constructed housing surround the modern city centres. In the rich neighbourhoods, state-of-the-art buildings exist alongside precarious shanty towns (these are known as favelas and are the result of land invasions). The socio-spatial provision of public services and facilities is extremely unequal. Public transport is insufficient and expensive and there is a poor quality of life.

As a result of intense speculation throughout the urban development process, the hoarding of state-serviced land has resulted in a large percentage of privately owned vacant plots in metropolitan areas, as well as marked discontinuities in the physical expansion of urban areas. This pattern of urban development has also produced a significant depletion of environmental resources and irreversible environmental damage. The result is the expansion of poor, low-density settlements over physically unsuitable areas and over areas with significant natural resources. The best-serviced areas, which are scarce and geographically concentrated, have been the subject of very intensive development, both horizontal and vertical. Put briefly, Brazilian cities, and especially the metropolitan areas, are inefficient and costly to manage, socially segregated, environmentally unfriendly and largely illegal.

Urban illegality and security of tenure

Urban illegality has taken diverse forms with the increased sophistication of the informal economy, which, in turn, has become particularly important in the process of urban land use and development. The failure to promote efficient official housing policies and the dynamics of largely speculative land markets have resulted in a scarcity of serviced land at affordable prices and in the lack of adequate housing options for the majority of the urban population. This has generated the proliferation of various forms of illegal land use and development and the widespread formation of illegal settlements. Security of tenure is largely absent and existing figures suggest that between 40 and 70 per cent of the urban population in the main cities have illegal access to urban land and housing opportunities. Recent data from São Paulo municipality show that at least 50 per cent (about 6 million people) of the total urban population live illegally; in other cities this can be as high as 70 per cent. These figures are expected to increase owing to overwhelming urban poverty.

There are historical links between secure land tenure, economic capacity and social mobility in Brazil. Therefore, the scale of urban illegality reveals the extent to which full socio-economic and legal political citizenship rights have long been denied. The vast majority of residents have not been fully recognized as legal residents of their cities, that is, as proper citizens, because they lack security of tenure. This lack has also made them politically vulnerable and financially incapacitated.

These growing illegal practices need to be urgently confronted because of their grave social, political, economic and environmental implications, not only for the immediately affected groups, but also for the broader social and urban structure. Urban studies have related socio-spatial exclusion with the growing fragmentation of urban space, generating costly and inefficient cities. More recently, links have been highlighted between socio-spatial exclusion (with growing urban poverty) and escalating urban violence (often related to drug trafficking networks). The importance of the informal urban economy and the variety of survival strategies of the poor are being acknowledged.

More than ever before, tensions between the formal and informal sectors, as well as those between the 'legal' and the 'illegal' cities, are becoming apparent. New social practices and new relations between the state and society are forged daily. The phenomenon of urban illegality is still under-researched and many analysts have yet to recognize that the apparent division between 'legal' and 'illegal' areas in cities is in fact closer to an intricate web of contradictory relations between official and unofficial rules and between formal and informal land markets. These 'illegal' social practices often enjoy greater social and political legitimacy than the legal ones. This has questioned the validity of the legal order.

The question of property rights

At this juncture, the role of the law and legal institutions in the process of urban development and socio-spatial exclusion must be critically discussed. As argued above, the exclusionary pattern of urban development in Brazil has been largely due to the nature of state intervention in urban development through its overall economic policies, lack of effective social and housing policies and distorted attempts at urban planning. It must be added, however, that the legal order, especially the anachronistic land-related legislation, has also played an important role in this process. Legal provisions and codes that do not reflect the socio-economic realities determining the conditions of access to urban land and housing, and elitist urban legislation, as much as the lack of land use regulation, have had a perverse role in aggravating, if not determining, the process of socio-spatial exclusion.

The central point to be addressed is that of the legal recognition of property rights, more specifically, urban real property rights. The traditional approach given to individual property rights, typical of classical liberal legalism, has long made possible the definition of real property merely as a commodity, thus favouring economic exchange values to the detriment of the principle of the social function of property. Given the historical context of unstable economic production, weak capital markets and chronic inflation, landownership has become a most important asset and one of the main capitalization mechanisms. Landownership has also become a surrogate for the country's inefficient social security system and therefore a significant factor of personal security and social mobility. It should be stressed that, as an inheritance from the centuries of colonial rule, most Brazilian land is privately owned, accounting also for the scarcity of public spaces.

Countless attempts at urban planning, land use and environmental control have been undermined by the existence of laws and/or jurisprudence, which greatly reduce the scope for state intervention (especially at the municipal level) in the field of individual property rights. The excessive hoarding of urban land has been tacitly encouraged by the individualistic nature of much of the legal order. The effective implementation of social housing policies, in turn, has been rendered difficult by (among other factors) high land market prices facilitated by the lack of proper regulation or by the exclusionary nature of rigid and elitist zoning laws.

Prior to the constitution, the legal protection and endorsement of private property rights was virtually absolute, leaving little room for state control over the use and development of urban property. In contrast, the constitution explicitly supported the principle of the 'social function of property and of the city'. Some important, socially oriented urban land and city management policies have since been undertaken in progressive municipalities, including pioneering experiences of participatory budgeting and land regularization programmes aiming at promoting security of tenure. However, as will be discussed, such progressive experiences have been constrained by the country's financial, political and legal realities. On the whole, urban legislation has rarely been concerned with democratizing land acquisition or land access for the poor; instead, urban laws contributed to the reproduction of segregation and the exclusion of the urban poor.

ACCESS TO URBAN LAND AND HOUSING

Given the regulatory and policy context described above, Brazil's urban and metropolitan structures and the prevailing conditions of access to urban land and low income housing have been determined by three main processes: first, the subdivision of central and peripheral areas (*loteamentos*); second, the invasion of land (favelas) usually in central areas; and third, the occupation of *cortiços*, or informal, unregulated rental accommodation usually in privately owned, poorly equipped houses or dilapidated buildings in decaying central areas. Favelas and *loteamentos*, both characterized by self-construction, have been the principal alternatives available to the urban poor. Alongside the acquisition of houses in the relatively few existing public housing projects targeting the disadvantaged population, they have provided housing opportunities for about 70 per cent of the urban population.

Throughout the whole process of urban growth, there has been an intense pressure on available land. The highly concentrated land structure has been gradually modified by several new practices of land use and development, generating new situations of tenure, as well as complex social property relations. The intensive rural–urban migration has brought about a growing need to urgently provide housing on a large scale, and several alternatives, both formal and informal, have been attempted.

The combination of rapid urbanization, sharp population growth, widespread expansion of economic activities and the scarcity of serviced land has redefined the role of urban land as a factor for investment. As mentioned above, an intense speculation process has resulted, generating all sorts of pressure for an increase in urban land value and the hoarding of vacant land. The overall process of urban development is reproduced in all of Brazil's main cities; it has been a process of urban spoliation (Kowarick, 1979).

However, there have been remarkable differences between the patterns of those *loteamentos* aimed at the middle and upper classes and those aimed at the less favoured segments of the population. While the former are usually located in more central areas, are technically modern and are recognized by law, the latter are usually located in peripheral areas and tend to ignore basic technical and legal requirements. Favelas remain the most precarious of urban settlements from both legal and technical viewpoints.

Irregular and clandestine loteamentos

A great number of *loteamentos* offering plots of land to the low-income population in peripheral areas have been developed irregularly, if not illegally, under precarious technical conditions. Their illegality is due to one or more of the following factors: first, the subdivision is not registered in the public registry office; second, the subdivision is developed in a rural area; third, the project does not obey the existing legal requirements; and fourth, developers fail to provide the promised urban infrastructure. *Loteamentos* developed in areas of contested ownership are popularly called 'clandestine'.

The process of expansion of peripheral areas through the development of irregular/clandestine *loteamentos* was well summarized by the 'nil economic cost versus infinite social cost' formula (dos Santos, 1980). The first stage in the development of most irregular *loteamentos* usually involves an improvised demarcation of some plots in a larger, undivided area, relatively far from the city centre, and sometimes within the rural zone: the area delimited by the municipal perimeter law where land subdivisions for urban activities are not allowed to take place. Such subdivisions are developed either by landowners, associate developers or even land usurpers, usually under extremely precarious and inadequate technical conditions.

This initial step is followed by the commercialization of the resulting plots, usually in attractive conditions; in many cases these first plots are given away free of charge for immediate occupation. After an incipient occupation, in the absence of any basic infrastructure (even roads), a subsequent stage of subdivision aims at selling the best plots at more expensive prices, and so on.

Between such successive stages, and subjected to increasing pressure from the new occupiers (and new voters), municipal administrations have had to gradually provide these *loteamentos* with basic infrastructure and services, thus contributing to an increase in the value of the whole area, particularly the vacant land. Municipal administrations have often been forced to change their perimeter laws in order to incorporate the newly urbanized areas into the urban zone, therefore partly legalizing the illegal settlements.

Housing in such *loteamentos* is self-constructed, frequently through *mutirão* (collective effort), usually with precarious materials and ignoring existing building regulations. The occupiers have bought their plots through a signed contract, from whoever presented themselves as landowners. In this sense, despite their irregular and/or illegal nature, they differ fundamentally from favelas, which originate from land invasion. However, with frequent problems concerning the original ownership of the divided land, the occupiers of *loteamentos* have often not been allowed to have their contracts registered at the registry office, as required by civil legislation.

Many *loteamentos* have been established on steep slopes, in areas subject to erosion, landslides or flooding, and without a proper road system and articulation with the neighbouring subdivisions. This has had a negative impact on the environment and created complex urban problems, the worst being the creation of an excessively large number of vacant areas for speculation, within and between land subdivisions. The main Brazilian cities are believed to contain 30 to 40 per cent serviced vacant land, technically urbanized/serviced through state action and therefore with public money. This multiplies the technical requirements and financial costs of service and infrastructure provision, making cities extremely difficult and expensive to administer. It exacerbates commuting distances and fragments communication between people and between places.

Favelas

Brazilian favelas, which result from the invasion of private or public land, are distinguished in legal terms from the above-mentioned clandestine or irregular *loteamentos*, in that favela dwellers, at least at the time of the original occupation, lack any form of property or tenure title. On average, 20 to 40 per cent of the population are living in favelas in the main cities, which have developed in tandem with the urbanization process.

Several studies have stressed how the invasion of private and public areas for the construction of favelas has been the most radical consequence of the process of production of the urban structure. Favelas are the socio-spatial result of a combination of historical factors involving both formal and informal socio-economic processes, the determination of the costs of urban industrial labour, as well as the process of land development and speculation. As they are also the product of state action in such intertwined processes, they have to be understood within the broader context of the failure of the national housing policy. Briefly, favelas are the crudest expression of the inequalities and contradictions long underlying the structure of Brazilian society (Fernandes, 1995).

Favelas were originally formed on land close to city centres. Owing to the high cost of public transport and the greater availability of services and jobs in inner-city areas, the invaded land became more valuable as cities grew. More recent favelas have been formed in peripheral areas in the main cities. Favelas differ in size, but some have a population of hundreds of thousands of people. They generally look the same: steep, hilly areas, densely occupied with a spontaneous, irregular and inarticulate pattern of land subdivision. The impoverished streets, alleys and stairways are confusing and not suitable for access or general traffic. Favelas lack almost every element of urban infrastructure and collective equipment and the precarious standard of most dwellings makes for unhealthy and dangerous daily living. The invaded land is largely unsuitable for human occupation due to geological and ecological factors.

Whatever the broader implications, the residents view favelas not as a problem, but as an indispensable solution to their housing needs, because both private urban developers and the state have failed to provide adequate alternatives. Many favelas have been established for over 100 years, with generations being born there. Some were originally the result of a collective invasion, others were less organized. Once a community is established, several forms of commercial transactions and civil relations guarantee the transfer of the plots and structures. Over time, better buildings have been built and the internal land markets have become more dynamic and complex.

LEGAL PROVISIONS AND THE STATE'S RESPONSES

However wide the implications of land subdivision may have been, the legal treatment of the matter has long proved unsatisfactory from the viewpoint of the needs of urban planning. State action has been essentially contradictory, treating differently the various processes of land subdivision. The state's responses to irregular/illegal subdivisions have ultimately depended on the issue of property rights.

Loteamentos

Until 1937, the only existing legal instrument in the field of real property, covering a few aspects of land use, was the 1916 civil code. However, the civil code made no reference to the matter of land subdivision for the commercial sale of plots. It contained some rules concerning construction and other minor aspects of land use, but the emphasis was on contracts and other related civil relations. Moreover, holding a highly individualistic conception of property rights, the civil code provided the state with very limited powers to interfere in the process of land use for reasons of social interest. Its few administrative restrictions concerned minor aspects of neighbourhood relations and, as such, it was unsuitable for the needs of urban planning (Fernandes, 1995).

Only in 1934 did the constitution adopt and promulgate the principle of the social function of property, aiming to make broader room for state control of land use and development. It was in this context that Decreto-lei (decree law)

No 58 was enacted in 1937. Its main goal was to guarantee the growing number of real estate buyers against irregular transactions and to prevent the proliferation of land subdivisions promoted by ill-intended developers. It revealed a deep concern with the question of the legality of property and, to prove their right to land, developers had to produce property titles going as far back as 20 consecutive years prior to the subdivision. There was also a detailed concern with the procedures necessary to enable the registration of new property titles.

However, Decreto-lei No 58 paid little attention to the urban dimension of the land subdivision process, failing to determine any minimal obligations (even the physical demarcation of the blocks, plots and streets) to be fulfilled by developers. Buyers were left without any legal guarantee that the projects' specifications would ever be implemented, apart from very limited civil (but not criminal) rights, mainly around compensation. While a programme of urban development in the corresponding plan had to be approved by the municipal administration, there were no specific requirements for the reservation of public, green and/or non-aedificandi areas in land subdivisions, other than streets and open spaces contained in the plan.

Though inadequate as an instrument of urban planning and land control, Decreto-lei No 58 was clearly a civil law concerned with formal relations between individuals involving land subdivision. Its enactment constituted an important development, as it was the first materialization of the constitutional precept of the social function of property. Put briefly, it aimed to put limits to the excessive contractual liberalism adopted by the 1916 civil code. Whatever its shortcomings, the legislation tried to provide a new outlook on the subject, and its ideological orientation was determined by the complex and contradictory political conditions then prevailing. While trying to modernize the legal institutional apparatus to prepare the cities for their new role as centres for industrial production, the new state had to fight for political hegemony, especially at local level, with the remnants of the formerly powerful rural oligarchies, the main landowners.

Decreto-lei No 58/1937, though dated, remained in force, unchanged, throughout the peak period of the urbanization process and was replaced only in 1979 when Federal Law No 6.766 was enacted. Furthermore, while the obsolete provisions of the 1916 civil code remain in force to date, the precept of the social function of property was never properly defined prior to the 1988 constitution. The liberal concept of property rights held by the civil code was always hegemonic and accepted as the dominant paradigm by the judiciary.

This means that most of the urban development process took place under very inadequate legislation, providing the municipalities with limited scope for interfering with the way private interests were promoting the subdivision of urban land. As there was no explicit constitutional reference to the matter of legislative power on urban development, the dominant jurisprudence affirmed that local laws could not restrain the scope of property rights, such as defined by the civil code.

Only in the mid-to-late 1970s did some municipalities succeed in enacting more progressive laws when, given the broader conditions of political opening, judges began to accept the notion that the legal regulation of urban land was indeed a matter for local peculiar interest and, therefore, within the scope of municipal autonomy. Thus, throughout most of the urban development process, municipal government was virtually excluded from the control of land subdivision, apart from the formal approval of plans and projects submitted by those developers who chose the legal route. Because of the prevailing legal definition of property rights, developers were allowed to predetermine, according to their own interests, the way they intended to subdivide the areas. The joke among frustrated planners was that 'green areas' in *loteamentos* were those areas on the plan painted green by developers and which, almost always, in reality corresponded to useless or eroded land.

Municipal administrations, towed away by private interests, would usually act *a posteriori* to minimize the problems created by the subdivisions. Municipal administrations could do little to force developers to regularize the irregular subdivisions and have them fulfil all the promises made in the 'programmes of urban development'. Generally speaking, the question of regularization was treated as a civil matter and, therefore, was restricted to the ambit of individual litigation between each buyer and developer.

However, the limited action of the municipal state was not without its contradictions and ambiguities, especially concerning clandestine *loteamentos*. Municipal administrations would lose no time before taxing the plots, even though they delayed the provision of services because of the illegality of the occupation.

The state has largely failed to complement the infrastructure supplied by developers, especially to guarantee the integration of the new subdivisions with the existing urbanized areas, to the socially unorganized and politically non-influential irregular *loteamentos*. The gradual upgrading of such areas only occurred from the late 1970s, as a result of the social mobilization around urban matters such as transport, water and electricity provision. Such mobilization brought about direct confrontation with the state and eventually led to changes in the overall political process.

However, if the occupiers of both irregular and clandestine *loteamentos* had to fight to have the state provide services and infrastructure, they at least had some basic security of tenure, that is, they were not expelled by official eviction policies (as has happened in favelas). After all, they had, for better or worse, bought their plots of land.

Federal Law No 6.766/1979 established general provisions governing the subdivision of urban land at national level, allowing federated states and municipalities to enact complementary rules to adapt the legislation to regional and local peculiarities. As a result, municipal government is now the true authority on the matter of urban land subdivision. It has legal conditions to lead the process, defining the criteria according to which such plans and projects have to be elaborated. In fact, administrations of cities with more than 50,000 inhabitants are obliged to define directives for the use of land in each particular subdivision proposal. This includes the demarcation of plots and streets, public areas and those areas reserved for collective equipment. It is aimed especially at integrating the new and existing divisions.

The new legislation completely changed the liberal orientation of Decreto-lei No 58 by establishing a series of technical requirements for the approval of *loteamentos*. Land subdivisions are no longer allowed in areas with serious safety, sanitary or geological problems, unless developers can counter the problems. There are several ecological provisions, including the reservation of many sorts of non-aedificandi areas. Moreover, it was determined that the plots cannot be less than 125 square metres, with the exception of specific urban developments and of subdivisions aimed at the construction of housing projects. As discussed below, this provision has been important for the formulation of policies for the regularization of favelas.

One of the most important moves was the requirement to reserve a minimum of 35 per cent of the developer's site for the implementation of urban services, collective facilities, a road system and other public open spaces. Law No 6.766 also turned these collectively used areas into public property on registration of the subdivision plan. For the first time, legislation stipulated a list of the minimal works to be carried out by developers. These may be extended further by municipal legislation, and comprise at least the development of the streets, the demarcation of plots, blocks and all public spaces and installation of a surface water drainage system, all within a two year period.

Law No 6.766 punishes with rigour all irregular activities relating to the subdivision of urban land. The legislation considers the development of any irregular subdivision, the sale of its plots and even the advertising of its existence, 'crimes against the public authorities'. It also proposes to make it easier for the state to promote the regularization of irregular *loteamentos*. Such regularization is to be done either by developers, when pressurized by buyers, or by the municipal administration, which would eventually be compensated by developers. Importantly, the new legislation extended the civil and criminal responsibility for the subdivision to any persons or groups who, together with developers, benefit from the undertaking.

Since its enactment several municipalities have adapted the content of Law No 6.766 to their local realities. They have conceived legislation on land subdivision together with zoning laws, allowing for an increasingly broad legal scope for state action and control. This was widened further by the constitution. On the whole, Law No 6.766 is an important, necessary and technically adequate legal instrument. However, it has one major flaw. The procedures it established for the regularization of irregular or illegal land may in practice only be applied to irregular *loteamentos* developed after its enactment. However, by 1979 most irregular land subdivisions, especially in the larger metropolitan areas, had long been consolidated, most plots had been fully paid for and in many cases it was impossible to identify the original developers.

Developers alone cannot share the responsibility for the regularization of long existing subdivision, as proposed by the legislation. Instead it has to be accepted as a duty of both the municipal administrations and the broader community through joint action between the public and the private sectors. Many administrations have tried to minimize their own investment by encouraging the practice of *mutirão* for the collective undertaking of works and construction of houses by neighbours. In many other cases, the municipal administrations have merely adopted a formal regularization policy (regularization by decree), neither undertaking the necessary works nor holding developers accountable. Such a policy has the immediate advantage of allowing the occupiers to register their plots and thus obtain legal security of tenure, but does nothing to solve, or at least minimize, the urban and environmental problems they face daily, nor the problems the irregular subdivision causes to the city as a whole.

As a result of the significant mobilization by progressive prosecutors for the government in São Paulo, Federal Law No 9.785/1999 recently modified Federal Law No 6.015/1973, which regulates the registration of land transactions. The new federal law makes the formal legalization of irregular *loteamentos* easier and thus provides better conditions for security of tenure to low income social groups who live in such peripheral areas. However, by also partly modifying Law No 6.766/1979, the conservative politicians at the national congress widened the original scope of the well-intended Law No 9.785/1999 in such a way that it abolished the legal requirement that at least 35 per cent of the site be reserved for public use (although municipalities can still impose this by local legislation). Several other legal provisions applicable to land subdivisions were relaxed, such as the requirement for basic urban infrastructure in 'social interest zones'. Such a discriminatory treatment of the urban poor, while facilitating the formal legalization of existing irregular *loteamentos*, may also encourage irregular developments, thus provoking further negative urban, environmental and social consequences. Again, local legislation can deal with the situation differently, but the fact is that most municipalities lack the technical, administrative and even political conditions to do so.

Favelas

Although favelas have long been an essential part of Brazilian cities, their relationship with the 'official city', particularly with the local state, has always been extremely difficult. An extensive literature has described how, up to the 1970s, intense conflicts and countless violent collective evictions were orchestrated throughout the country. Because of the increasing mobilization of favela dwellers between the 1970s and mid-1980s, aided by sections of the Catholic Church, public policies were progressively forced to favour the relocation of the favela population. However, this was largely unsuccessful, mainly because of popular dislike of the poor housing alternatives usually offered by the local state.

Until recently, the issue of the rights of favela dwellers was, in strict legal terms, straightforward. As invaders, favela dwellers had no rights, since there was little room in existing legal principles for any socio-economic or political justification of such a radical infringement of property rights (Fernandes, 1995; Fernandes and Rolnik, 1998). The defence of landowners' rights was virtually unconditional. To date, the 1916 civil code and the civil procedure code grant both private landowners and the state several legal instruments with which to evict the invaders. Before the constitution, apart from property expropriation by the state through the payment of compensation, the only possibility for prescriptive transfer of ownership rights was *usucapião*, a form of adverse possession.

According to the 1916 civil code, the right of *usucapião* requires, in most cases, a pacific, uncontested permanence on the land for 20 years. It implies the loss of property due to non-use by the original owner and, as such, requires no financial compensation from either the occupier or the state. However, it provides landowners with a range of legal mechanisms to question, in court, the occupation of the land. It is virtually impossible to apply, because the high mobility and collective nature of the settlement process does not conform to the technical and individualistic requirements of civil legislation. Moreover, the right of *usucapião* is not applicable to invasions of state-owned areas and, therefore, it cannot be claimed by approximately 50 per cent of all favela dwellers who have occupied public land. Several constitutional and legal provisions prevent the configuration of *usucapião* in public areas, for example, article 183 of the constitution and article 200 of Decreto-lei No 9.769/1946.

In keeping with the individualistic ideology of the civil code, all constitutions prior to that of 1988 established that only federal legislation could govern property relations, as these would have a civil – and not an urban or social – nature. Given that Brazil to date has no national legislation on land use and development, no proper national planning system and no clearly defined national urban development policy, the provisions of the civil code have long restricted the ambit of state action. Modified several times since the beginning of the 1980s, the current version of the bill on urban development (national law) is the so-called 'City Statute', which has not yet been approved by the national congress.

This limitation was felt even more in the field of favelas, where any daring initiatives faced strong resistance from the conservative judiciary. It is true that, since the mid-1930s, the traditional concept of property rights has been increasingly challenged by several urban laws and some judicial decisions, which have attempted to materialize the principle of the social function of property. However, although such progressive judicial decisions recognized and increasingly broadened the rights of tenants and other people in irregular occupation, until recently they failed to include favela dwellers, who were legally viewed as mere usurpers – land invaders lacking formal titles or contracts.

It is only in the last two decades that some municipal authorities have started to recognize that favela dwellers have a right of access to urban land and housing. Several programmes have already been formulated aimed at both physically upgrading and legalizing existing favelas. Most programmes have been implemented at the municipal level, although some important policies have also been formulated at the federated state level. The first significant programmes of land regularization were undertaken in Belo Horizonte and in Recife in the early 1980s. These were a direct result of the growing socio-political mobilization in these cities. These programmes had as main legal grounds the provisions of the then newly approved, above-mentioned Federal Law No 6.766 of 1979, which governs the subdivision of urban land nationally. This law created the vague concept of 'specific urbanization', implying the possibility of partly treating some specific situations of land subdivision with criteria different from the general ones (Fernandes, 1995).

Belo Horizonte's zoning law was originally enacted in the mid-1970s, constituting a pioneering legal instrument for municipal control over land use, subdivision and development. According to its new zoning law of 1976, some areas which should be subject to special forms of urbanization (such as preservation areas and areas reserved for public utilities, etc) were initially classified as 'special sectors 1, 2 and 3'. Making use of the concept of 'specific urbanization', the zoning law in the early 1980s was altered and local favelas were classified as 'special sectors 4', in an innovative, though still tentative, legal formulation. Prior to the creation of the special zones, favelas had been explicitly ignored within the scope of the existing municipal zoning, repeating somehow the same situation created by Belo Horizonte's original master plan, in which no space was allocated to the workers who built the city. As a result, in 1895, even before the city's inauguration in 1897, about 3000 people already lived in favelas (Fernandes, 1995). Many other municipalities have since copied this formula, with the current technical terminology being 'AEIS–special areas of social interest' or 'ZEIS–special zones of social interest'.

The new laws in Belo Horizonte and Recife constituted a remarkable advance in the recognition of social rights in Brazil. They contributed to the characterization of favela dwellers as people with rights, and introduced the outlook that, as citizens, favela dwellers should be granted a place – and a space – in the city. The eventual incorporation of favelas into the existing municipal zoning was indeed the most appropriate move if favelas were to be treated within the context of the city's urban structure. However, as could be expected, the first zoning laws and favela legislation were the object of fierce legal criticism. This was translated into conservative judicial decisions, on the grounds that the civil and constitutional legislation in force did not permit such daring state intervention in the field of private property, particularly by municipal government.

Interests at Stake, Social Practices and Legal Pluralism

From the developers' viewpoint, the undertaking of *loteamentos*, especially those for the low-income population, has long been an attractive source of easy profit making. Developers have counted on the state's actions in the provision of infrastructure, therefore increasing the value of the land they keep vacant. The great number of subdivisions undertaken illegally have further reduced the investment of capital on the part of developers.

The political struggle for hegemony at the municipal level, combined with the instability of economic production since the 1930s, has made room for an increasingly significant speculation process. Because of the highly concentrated land structure, the growing land market has been increasingly divided between landowners, on the one hand, and urban developers, promoters and builders, on the other. While the first group has basically been interested in speculation, because of the scarcity of land, the latter groups have been gradually affected by the high costs of land acquisition. The tensions between such groups have increased.

Although property developers have been the main beneficiaries of the whole land subdivision process, several other parties have gained from it to differing extents. From the perspective of most urban dwellers, especially the mass of migrant workers coming from the countryside, the acquisition of plots in *loteamentos* alongside favelas has constituted the most accessible solution to their urgent housing needs. Favelas and *loteamentos* are a cheaper, more informal and more immediate means of access to housing than those offered through the official housing mechanisms (which have been appropriated by the middle classes).

Favelas and *loteamentos* have constituted different, though not exclusive, stages of a broader social process. After having initially spent some time in central favelas, a large number of people have eventually bought a plot of land, usually in a more peripheral area. This has not only been due to the difficulties of settling down in invaded areas, but also because most people see the ownership of land as an important factor of social stability and personal fulfilment. However, from a broad economic perspective, the reality is different. The intensive practice of *loteamentos*, once taken hold of by the speculation process, became a most significant example of the widespread practice of the privatization of gain and socialization of loss.

The process of intensive urban growth through the expansion of peripheral areas has had two principal consequences. First, at the social level, making use of several means of political pressure, local dwellers have gradually managed to invert, to some extent, the 'nil economic cost versus infinite social cost' relation. Through complex and contradictory political alliances with developers, local dwellers have managed to force the state to provide collective infrastructure and services, thus improving the conditions of daily life. In a perverse way the state has also made political gains, by perpetuating clientelist-type links with the local population.

Second, at the level of urban structure the problems become even more serious, revealing the impossibility of such an inversion of costs. In the context of the city, land subdivisions are neither isolated nor independent. Instead of solving the problem, the attempts at regularization through state action have often resulted in the creation of new peripheral areas – where the same is repeated – making the original urban area expand boundlessly and making it even more difficult for the state to overcome the social costs. Furthermore, the inversion of costs becomes impossible because of both the excessive hoarding of vacant land and the low density of the occupation of new urban subdivisions.

The action of urban legislation

The lack of an adequate legal treatment of this problem has facilitated the proliferation of irregular/illegal land subdivisions throughout the peak of the urbanization process. The maintenance of the provisions of the 1937 Decreto-lei No 58 was fundamental in maximizing productive and speculative gains in the real estate sector. This was instrumental in keeping low the costs of reproduction of the new labour force (especially concerning housing costs). It was also significant in the broader process of reproduction of the urban social relations of production.

The enactment of the progressive Federal Law No 6.766 of 1979 has to be explained by the combination of three major factors. First, by the mid-1970s the growing social mobilization process around urban matters was already putting the state's legitimacy at stake. In São Paulo, specifically, intense social mobilization called for the regularization of irregular/clandestine *loteamentos*. Law No 6.766 did recognize such subdivisions to some extent, but failed to provide adequate instruments to enable their effective regularization. Second, the enactment of Law No 6.766 was a strategic gain for the national council for urban development, the then newly created planning agency. This agency needed to carve out a politico-institutional space for itself, in a context of growing resistance (on the part of the municipal administrations) from whatever attempts at urban intervention were made by the authoritarian federal government. Third, and perhaps more important, the proposed law was approved by the national congress, without major amendments, because of support from a coalition of real estate parties. Support came from developers and builders from São Paulo, whose interests were being undermined by the uncontrolled dynamics of the land markets.

Another significant phenomenon since the enactment of Law No 6.766 has, ironically, been the reinforcement of the process of social segregation in space. This resulted from increasing state action towards the enactment of legislation, as well as the implementation of effective policies of land use control in the main cities, in combination with the logic and mechanisms of the land market.

Many municipal administrations have attempted to interfere in the existing urban structure either directly (through policies of urban renewal and the regularization of favelas) or indirectly (through the enactment of urban and fiscal legislation). However, since the state has failed to control the speculation process, the 'cleaning' of central areas has often implied the expulsion of the urban poor to the outskirts of the cities. The same process has happened when state intervention eventually reaches the peripheral areas closest to the city centre, and so forth.

State action, through the enactment of urban legislation, has become more significant in determining the rates of profit in the land market. Being directly linked with the production of economic space, urban legislation plays an increasing role in determining the cost of land and housing. However, the social dimension of such a process remains relegated. While Law No 6.766 failed to provide instruments to enable the effective regularization of pre-existing irregular and illegal *loteamentos*, the pressure exerted by the powerful speculation process has added to the financial burdens brought onto the occupiers by the enactment of urban and tax laws.

The same social costs have resulted from attempts to regularize favelas. While successful to varying extents in terms of physical upgrading, they have failed, on the whole, to properly address the popular claim for the legalization of the plots. Ironically, while these upgraded favelas are better provided with infrastructure and services than many peripheral *loteamentos*, the latter have been more easily legalized. Most laws for the regularization of favelas have recognized some form of property rights for favela dwellers. However, it appears that the recognition of land invaders as owners remains a threat to the state and to public opinion influenced by the ideology of property rights. Most experiences have been based on the expropriation of the land, with the state financially compensating the previous landowners. This signals that the invaders have no rights of their own.

The contradictory stance of the state towards such a radical solution has frequently meant ineffective enforcement of progressive legislation. Yet, in many cases, the symbolic concessions seem to have disarmed the favela movement and integrated much of its leadership, just at the moment when a renewed mobilization would have been crucial. The ultimate irony is that, far from subverting the traditional order, legalization of favelas risks becoming another factor in the process of social segregation, leading to the expulsion of the population to more precarious settlements (Fernandes, 2000).

State action and legal pluralism

The state's refusal to confront the issue of land regularization has had several implications. It has led to a virtual rupture in the relationship between the state and the population living in irregular areas, especially in favelas. Throughout the urbanization process the existing legislation has increasingly kept a distance from the social reality of everyday life. The immense population, excluded from regular access to land and from the overall decision-making process, has had to create its own channels of participation, and new forms of collective organization, interest association and leadership have emerged. With the generalized discrediting of traditional forms of political representation, such new socio-political agents have been gradually legitimated by the population.

Another related phenomenon has been the formation, by the population segregated in irregular *loteamentos* and especially in favelas, of a type of 'informal justice' or alternative legal order. Complex civil and commercial codes have long been developed, reflecting the dynamic social relations. They have dealt with diverse issues ranging from: rent, sale and donation of plots and structures; marital and inheritance rights; subdivision of land and rights of passage; to installation, functioning and transfer of commercial sites and activities.

More recently, in those favelas where organized drug trafficking activities have dominated, the formation of an incipient criminal code has also been identified. Local dealers – owners of the territory – have applied distorted principles of popular justice in cases of burglary, robbery, rape, financial debts, 'moral' misbehaviour, and so on. The reported punishments have involved humiliation and several forms of corporal punishment, as well as capital punishment for the gravest of all crimes, the disrespect for the trafficking monopoly in each territory. Dealers have been reported to invest in the upgrading and security of the irregular areas, as well as in the improvement of the local community, and some local leaders seem to have been increasingly legitimated by the population. However, even the patterns of illegality have somehow been determined by the official legal system in that, aiming at being accepted as legitimate, the unofficial rules have incorporated, and have been structured around, the fundamental prevailing legal principles. In other words, the existing forms of legal pluralism have been constituted in a conflicting dialogue with state law.

Given this conflict between the legal and the informal orders, that is, between situations recognized by state law and those that have been given legitimacy by popular struggles, even Brazilian cultural imagery has been partly modified, gradually incorporating new elements into the traditional concept of private property, principally the idea of the social function of property. Throughout the century, the favela movement has gradually conferred legitimacy on what had long been dismissed as a marginal claim, that for the right of tenure. Without the recognition of such a claim by official legislation, the core of the legal order would eventually be exposed as illegitimate.

Despite the liberal idea that the concept of private property is universal and all encompassing, in reality in Brazil there has always been a differential treatment of the various social relations established around the use, occupation and development of urban land, including those involving some form of illegality. The fact is that the state, and society for that matter, responds differently to different situations, as if there were 'degrees' of illegality, some more acceptable and tolerable than others. Throughout the decades of intensive urbanization, the state traditionally reacted in different ways to the illegality resulting from the three main forms of land subdivisions.

First, while keeping taxation low for reasons of political convenience, it has provided all the infrastructure works neglected by the developers of authorized *loteamentos* for the middle and upper classes. It has also accepted the enclosure of luxury *loteamentos*, despite constitutional and legal provisions that forbid the development of urban activities in rural areas, the prevention of access to the coast and the disruption of traffic in public streets.

Second, it has tolerated the existence of irregular or illegal peripheral *loteamentos* for the low-income population; while delaying the undertaking of works and the provision of services, the state has often taxed the population. It has failed to hold developers responsible for the irregularity. However, it has also ambiguously admitted the occupiers' rights to stay on the land they have bought. When pressurized, the state has often recognized the subdivisions, though applying technical criteria different from, and inferior to, those applicable to the 'legal' city.

Third, the state has long refused to recognize the rights of favela dwellers who have been subjected to violent eviction, expulsion and removal policies.

It could be argued that the existence of property titles or proof of tenure should result in decisive state responses. Official policies, laws and judicial decisions conferred some degree, or at least an appearance, of legitimacy on the processes by which social agents have dominated the access to land. However, the legal political process of legitimization has not corresponded to the concrete forms of land appropriation.

When it comes to the case of invasions, it has been particularly difficult for the state to define a compromise formula to enable the legitimization of direct land appropriation. The problem remains how to turn the de facto situation into a socially acceptable legal form of landownership, one that reflects the corresponding social process. Since the very pillar of the socio-economic process is clearly at stake, without any deceptive ideological mask, the state has to be careful of the consequences, both material and symbolic, of its actions. A very long time has usually separated the beginning of the regularization process from full regularization. During this period, the state has gradually had to work out the necessary social form of legitimization. The enactment of progressive legislation – without the real intention of enforcing it – seems to be the latest stage of this process.

However, there is an urgent lesson to be learnt from the repeated violent episodes in the favelas of Rio de Janeiro and São Paulo. They suggest that the relation between the state and the segregated population seems at present to be threatened by a new factor. Traditional social mobilization in favelas, usually structured around the claim for property rights and urban development, seems to have come to a standstill as a result of the lack of effective public response, including, in some cases, the non-enforcement of the legislation promising the long claimed access to property rights. The crisis in favela leadership that has followed this demobilization may have left not only favela dwellers, but also the state, without their main interlocutor in the process of negotiation and social control. The contradictory political history of favelas has shown that the state should not take the existing, though precarious, legitimization pact for granted; the pressure for making ends meet in their daily lives makes favela dwellers highly unreliable political agents. Life in the favelas continues, and new forms of leadership, rules, alliances and forms of collective action have been established around issues other than property – but this time outside the reach of state action.

URBAN POLICY IN THE 1988 FEDERAL CONSTITUTION

Before the constitution, the state had little scope for intervention in urban areas, other than that determined by its elitist, undemocratic nature. In the context of Brazil's contradictory federal system, the lack of a constitutional treatment of urban and territorial jurisdiction issues led to endless legal controversies and conflicts between federal, federated state and municipal administrations. These concerned the power to enact urban legislation and implement urban policies. State intervention in urban areas, particularly during the years of authoritarian rule between the 1964 military coup and the constitution, was largely decided at the federal and federated state levels, under conditions of weak political and fiscal accountability. This corresponded with attrition of the autonomy of more than 5000 municipalities, which lacked legal, technical and financial resources and instruments to tackle the problems brought about by intensive urbanization. Moreover, the urban population was virtually excluded from the law and decision-making processes on urban questions at all levels, especially in the nine institutionalized metropolitan regions, which were administered in a blatantly authoritarian fashion between 1973 and 1988 (Fernandes, 1995).

It was in this context of distorted political institutional planning that the private interests of minority groups led the process of urban development, towing the state along in their search to facilitate the conditions of capital accumulation in cities. Most state interventions consisted of sectoral policies and, as such, they were capitalized on by the economic groups that controlled the state; whereas the needs of the urban poor were neglected. Urban space in Brazil clearly reflects, and reinforces, the unequal conditions of wealth distribution in the country (among the world's worst). It also reflects and reinforces the tradition of political exclusion of the less favoured population.

This contradictory and conflicting process is directly related to the broader process, which has promoted changes in Brazil's socio-economic and political order over the past 20 years. Since the late 1970s, federal government enacted more progressive urban legislation, such as the laws governing the subdivision of urban land and the formulation of a national environmental policy. On the one hand, this was a response to the increasing social mobilization and, on the other hand, it was a response to pressure by important sectors of the real estate capital to create new opportunities for capital investment and reproduction in urban areas.

The pace of legal political reform has been slow, given the complexity of the diverging social, political and economic interests involved. While many different versions of a national urban reform bill have been inconclusively discussed at the national congress since 1983, the scope for municipal action under general national directives has, on the whole, been widened. Of special importance was the promulgation of the 1988 federal constitution. Its innovative constitutional chapter on urban policy and its improvement to the conditions for the political participation of the urban population in the law and decision-making processes, resulted from a process of intensive social mobilization.

Various organizations of civil society formulated and submitted to the constitutional congress the remarkable 'popular amendment on urban reform'. This defined the notion of social property in such a manner that it would impose itself as a new paradigm, replacing the one established by the 1916 civil code. The main claims put forward by the popular amendment were: the recognition, regularization and upgrading of informal settlements; democratization of access to urban land and measures to combat property speculation; and the adoption of a democratic and participatory form of urban management. Several compromises were eventually agreed upon and an innovative chapter dedicated to urban policy was eventually inserted in the constitution.

The constitutional provisions

It was only through the 1988 federal constitution that the general legal provisions on the matters of urban policy and property rights were considerably improved. It contains a more coherent approach to the urban development process as a whole, defining the notion of the 'social function of property' as the very condition for the recognition of individual property rights. It also improved the conditions of legal, political and financial autonomy of the local state (Fernandes and Rolnik, 1998).

While the constitution makes no specific mention of favelas, it considers the provision of housing as a matter for the concurrent power of the federal union, federated states and municipalities. Each should 'promote housing construction programmes and the improvement of the existing conditions of housing and basic sanitation'. They should also 'combat the causes of poverty and the factors of marginalization, promoting the social integration of the less favoured sectors' (1988 federal constitution, article 23, IX and X).

Three major points deserve to be stressed. First, the constitution conferred on the municipal authorities the power for the enactment of laws governing the use and development of urban space, in order to guarantee the 'full development of the city's social functions' and the 'welfare of its inhabitants' (article 182). Cities with more than 20,000 inhabitants are now obliged to approve a master plan law, which is considered to be the 'basic instrument for the development and urban expansion policy' (paragraph 1).

Second, the right of private property was again recognized as a basic principle of the economic order, provided that it accomplishes a social function according to the 'dictates of social justice' (article 5, XXII, XXIII and article 170, II, III). Significantly, however, it was stated that urban property only accomplishes its social function when it attends to the 'fundamental requirements of city orderliness expressed in the master plan' (article 182, paragraph 2). As a result, the more developed and comprehensive the master plan, the more advanced and progressive the conception of social property. Rather than having a predetermined content, the property right is supposed to turn into a 'right to property', a socially oriented 'obligation'.

Third, another progressive development was the approval of the right of special urban *usucapião* for those who have occupied less than 250 square metres of private (never public) land for five consecutive years (1988 federal constitution, article 183). Proposed with the actual situation of favela dwellers in mind, this aimed to render regularization policies more viable. However, it has not yet been regulated by ordinary federal legislation and, again, the possibility of its full enforcement is questionable. Nevertheless, applied in theory to perhaps half of all existing favelas, it is a major step towards recognizing favela dwellers as citizens.

There were two important innovations that may contribute to changing the quality of the city planning process. First, the constitution reserved for the municipalities the most important role in this process. Although both the federal and federated state governments have concurrent power to enact laws and to formulate programmes on the matter of land use, their scope is limited to very generic directives or related to specific situations that cannot be solved at local levels. Municipal government, responsible for the enactment of urban legislation and for the implementation of urban policies, is the real authority on the subject.

Taken alone, however, this development was not enough to change the character of the city planning process. It was clear from past experience that, to be effective, the handling of urban questions presupposes the democratization of decision-making. In general, the processes of political administration and conflict resolution, which have determined the models and techniques of plans and the choice of instruments of control, have not resulted from a social process with significant participation of the different interests in land use. It is important that the constitution ensured the possibility of some, though still very incipient, degree of popular participation in the decision-making process of urban questions. Besides favouring the reinvigoration of classical representative democracy, the constitution accepted the possibility of relatively direct participation at local level in the urban planning process. Non-governmental organizations are now able to formulate bills on some urban matters, and to submit them to the municipal legislature.

The very existence of this precept is a sign of change in the political process. Moreover, it coexists with a broader recognition of the 'collective right for diffuse interests' (article 5, LXXIII), which offers more scope for citizens to have recourse to judicial power for the defence of urban rules, as well as of social and environmental values.

Therefore, the economic content of urban property is to be decided by municipal government through a participatory legislative process, and no longer by the exclusive individual interests of the owner. This principle was also realized through the creation of new instruments such as compulsory edification, progressive taxation and flexible expropriation, which, together with other instruments to be created by local legislation, aim to put the local state in the lead of the urban development process. The local population is now entitled to participate in decision-making over the urban order, both through their elected representatives and, directly, through the action of urban non-government organizations (NGOs).

The constitution recognized that the decision-making process of urban questions is indeed a political process, to which the definition of the patterns and limits of economic exploitation of real property is entrusted. For the first time, the population was to some extent considered to be a political agent. Therefore, popular mobilization against dominant economic groups is now expected to take place, also within legal and institutional spheres. While the state was confirmed as the preferential promoter of the urban growth process, a new social right was also recognized, 'the right to urban planning'. Much more than a mere faculty of the municipal administration, it is one of its main legal obligations, as well as an expression of social citizenship. The main novelty, surely, is that urban law has unequivocally been put where it has always belonged, namely, in the political process.

Whereas the main prevailing difficulties preventing the recognition of security of tenure in irregular/clandestine *loteamentos* have a financial nature, one would expect that all legal controversies on favela legislation had ended with the inclusion of the 'social function of property and of the city' in the constitution. However, while there is decreasing resistance to the approval of general zoning laws in most cities, those laws supporting favela regularization programmes still face fierce opposition, especially when it comes to the legalization of the invaded plots.

Nevertheless, the effective enforcement of a growing number of urban and environmental laws, especially at municipal level, has been seriously undermined by the perseverance of several political, financial, institutional and legal factors. The main legal constraints to proper law enforcement are the archaic structure of the judicial system, the problematic definition of the federal regime and the overall organization of the political system. Though greatly improved by the constitution, the legal order in Brazil still does not express the nature of the existing urban territorial order. Nor does the political institutional order, recognized by the constitution, express the real dynamics of socio-political relations. The distortions provoked by the legal institutional order have been instrumental in maintaining and widening long-existing economic inequalities and social injustice.

Prospects for progressive urban planning and participatory urban management

Following the trail of the constitution, several federated-state constitutions and municipal organization laws have recently reinforced the above-mentioned progressive principles. Now that municipalities are stronger politically and somewhat richer, many interesting experiences of urban administration have been attempted. While the prospects for urban planning are better than ever before, the scale of the urban, social and fiscal crisis formed throughout the last four decades is enormous. Brazil is still struggling with a deep economic and fiscal monetary crisis; therefore the reach of such local experiences has been limited by the broader national context. Since the 1980s, metropolitan areas have been badly affected by the crisis. Both the increase in poverty and escalating urban violence have contributed to further deterioration in already low living standards and in the fragmented social fabric. The changing local political realities have also been decisive as to the success or failure of attempts at comprehensive planning.

Whatever its shortcomings, urban reform is gradually taking place, supported by an increasing social awareness of the urban question. Nevertheless, the changes in legislation brought about by the constitution were certainly not sufficient to structurally modify the previous situation. Much of the excluding order of distribution and exercise of political power was maintained. Urban policies and the standard of living of urban society will only change if the political process in the country leads to deeper changes in the class-bound nature of the state, hence improving the conditions of legal, political and socio-economic citizenship of the Brazilian people. In this context, the future of urban planning and the fate of cities will depend on how the current political reforms combine measures of wealth distribution with the recognition of citizenship rights of urban dwellers. Therefore, the social mobilization of the urban population is of crucial importance. The success of local initiatives will primarily depend on the consolidation of the country's fledgling democracy concerning the promotion of social justice.

Indeed, despite the improvement of the legal order, the reality is that the vast majority of the population in the larger Brazilian cities still live in irregular, and very often illegal, conditions. While new invasions occur almost daily, the progress of the legislation has not yet altered the situation of most long-consolidated favelas and irregular or clandestine *loteamentos*. The authoritarian distinction between the 'legal' and the 'illegal' cities remains and has even been reinforced by the perverse impact of the new laws on the speculative land market.

The constitution adopted a broader approach to both diffuse interests and consumers' rights, providing some mechanisms, which, if combined, may prove powerful for the judicial defence of collective interests in urban matters. In this context, the change in the current judicial attitudes to the matter of state intervention in the definition of property rights will depend on how successfully social organizations, especially those representing the interests of the vast population living in irregular or clandestine *loteamentos* and favelas, manage to put their claims forward.

The question of access to urban land and housing cannot be left to market forces alone, and the recognition of the long-claimed 'right to the city' is the condition for the consolidation of Brazilian democracy. However, urban reform is not given by law, instead it has to be attained through the political process (Fernandes, 1995).

It is in this highly political, complex context that the following chapters should be understood.

Conclusion

Political redemocratization and economic restructuring have recently had several implications for land policy in Brazil, including the approval of legal, urban and fiscal measures aimed at addressing, to some extent at least, the need to provide affordable access to land and housing and security of tenure for the urban population. In fact, a fundamental change of orientation of the public policies regarding illegal urban settlements has become evident in many cities.

After decades of removals from illegal settlements, the state has become increasingly tolerant. Eventually, though reluctantly, the state has proposed the legal and technical regularization of illegal settlements. Since the 1980s, partly as a result of the increasing social mobilization for the recognition of housing rights, several regularization policies have been formulated and implemented. They had the explicit objective of promoting the socio-spatial integration of the excluded communities living in favelas and peripheral *loteamentos*. In the absence of a national policy, local experiences, combining physical upgrading and legalization of the illegal settlements, have varied since the pioneering experiences of Belo Horizonte and Recife in the mid-1980s. With the 1988 federal constitution, they have gained a new vigour. However, they also suffer the effects of legal, political and financial constraints on progressive urban management in Brazil.

Despite efforts by the collective entities of the national forum for urban reform, and regardless of the recent development of Brazilian urban legislation at the municipal level, most regularization policies are still unsatisfactory with regard to the legalization of the occupied areas, particularly those illegal settlements that occupy privately owned urban land (Fernandes, 2000). The necessary political conditions have not yet been created to promote the reform of liberal legalism still prevailing through the paradigmatic 1916 civil code, which is still in force and does not recognize the centrality of the principle of the social function of urban property.

It should be stressed that reforming the national legal political order is only part of the challenge. The concentrated and exclusionary land structure formed throughout centuries of economic development in Brazil must be urgently confronted, both in cities and in the countryside, as an indispensable condition for the promotion of regularization policies. Urban reform cannot be dissociated from agrarian reform, as emphasized by the critical experiences of several countries in Africa and in Latin America, including Brazil.

As a result of increasing social poverty, even the acquisition of cheap plots in irregular *loteamentos* is beyond the possibilities of an increasingly large group of urban poor, and the 'problem' of favelas is worsening. Besides the increase in residential densities in traditional favelas, new favelas have formed on the urban peripheries. Private and public areas have been invaded daily, mainly in the surrounding cities of the metropolitan areas, where legal control of land development has not been as strict as that applying to central areas. Political demobilization and new socio-economic activities, including those revolving around increasingly organized drug trafficking, seem to be transforming life in the favelas. Notwithstanding government involvement in some basic upgrading works and public services in the main favelas, and to a lesser extent in peripheral *loteamentos*, on the whole state action has failed to promote the integration of the favela areas and irregular *loteamentos* into the broader socio-spatial urban context. Thus the prospects for traditional regularization programmes are not very promising, especially as far as the objective of socio-spatial integration is concerned (Fernandes, 2000).

Past experience has shown that the approval of legislation does not guarantee its enforcement. Favela dwellers, as well as residents in peripheral *loteamentos*, must redefine their means of collective action in order to have regularization measures implemented. Having fought long for the law, they must keep on fighting to see it materialize. In the case of favelas, the question of perceived security of tenure must be addressed. More local programmes involving the relocation of favela populations have been adopted in recent years.

It has to be stressed that regularization policies have an inherent remedial nature. They should not be dissociated from much-needed comprehensive public policies based on direct state intervention and substantial public investment in urban areas through, for instance, large-scale rehabilitation projects, social housing and urban renewal programmes. They should be based on effective preventive policies aimed at transforming the nature and dynamics of urban development to the benefit of the whole community.

In São Paulo, an important process of social mobilization has led to 'Projeto Moradia' (the housing project). This proposes the improvement of housing conditions in existing *cortiços* and the residential occupation of the great number of vacant buildings in the city's central areas. To be successful, this and other housing strategies would require explicit political and financial commitment on the part of the local authorities.

The recognition of some form of land tenure rights of the population already living in illegal and precarious conditions (*cortiços*, favelas and *loteamentos*) is of utmost importance. Whatever the chosen legal formula, it is fundamental that the recognition of urban and tenure rights takes place within the broader, integrated and multisectoral scope of city (and land use) planning, in order to prevent distortions in the land market and thus minimize the risk of displacement of the traditional occupiers. The laws supporting regularization programmes must be properly integrated into the overall urban legislation in a given context. Moreover, for a legal solution to work properly, it must result from a democratic and transparent decision-making process that effectively incorporates the affected communities.

Above all, the recognition of housing rights and of security of tenure must be promoted within a broader context in which urban reform and law reform are reconciled. Law reform is a direct function of urban governance. It requires new strategies of urban management based on new relations between the state (especially at local level) and society, renewed intergovernmental relations and the adoption of new forms of partnership between the public and the private sectors, within a clearly defined legal political framework. Law reform requires a fundamental renovation of the overall decision-making process, so that traditional mechanisms of representative democracy and new forms of direct participation are combined. The democratization of access to land and housing depends on the democratization of the political process as a whole. A most promising experience is that of 'participatory budgeting' adopted in several Brazilian cities, in which significant scope has been created for the participation of community-based organizations in the formulation of the local investment budgets.

Finally, the need to promote comprehensive legal reform and judicial review can no longer be neglected, especially in order to promote the recognition of collective rights and to broaden collective access to courts to guarantee law enforcement. To some extent, Brazil has already incorporated the notion of collective rights in its legal system, thus enabling the judicial defence of so-called diffuse interests in environmental and urban matters by both individuals and NGOs. Put briefly, urban reform and the recognition of security of tenure are not to be attained merely through law, but through a political process that supports the recognition of the long-claimed 'right to the city', not only as a political notion, but also as a legal one. The collective action of NGOs, social movements, national and international organizations, and individuals within and without the state apparatus is of utmost importance to guarantee both the enactment of socially oriented laws and, more importantly, their enforcement. There is also a fundamental role to be played in this process by lawyers, judges and prosecutors for the government.

If these are truly democratic times, the age of rights has to be also the age of the enforcement of rights, especially collective rights. It is only through a participatory process that law can become an important political arena to promote spatial integration, social justice and sustainable development.

References

dos Santos, C N F (1980) 'Velhas novidades nos modos de urbanizacao brasileiros', in L Valladares (ed) *Habitacao em Questao*, Zahar, Rio de Janeiro

Fernandes, E (1995) *Law and Urban Change in Brazil*, Aldershot, Avebury, UK

Fernandes, E (2000) 'The Legalization of Favelas in Brazil: Problems and Prospects', *Third World Planning Review* Vol 22, No 2, pp167–187

Fernandes, E and Rolnik, R (1998) 'Law and Urban Change in Brazil', in E Fernandes and A Varley (eds) *Illegal Cities: Law and Urban Change in Developing Countries*, Zed Books, London and New York, pp140–156

Kowarick, L (1979) *A Espoliacao Urbana*, Paz e Terra, Rio de Janeiro

Chapter 7

Security of Tenure in São Paulo

Ellade Imparato

There is widespread poverty in Brazil. Approximately 60 per cent of the population earns less than US$100 per month. They work in the informal sector and live in informal housing. This means that social exclusion affects 6 million people in São Paulo, Brazil's wealthiest city. Security of tenure is one of the tools that can be used to assist in the eradication of poverty.

The legal apparatus to secure tenure in informal settlements is in place, but local governments need to be pushed to establish broad programmes. Without deep municipal involvement, 'adverse possession' is not an efficient policy. The involvement of municipal government is needed, not only because it is closer to the informal communities, but also because it has a duty under article 182 of the 1988 federal constitution.

The country's urbanization process occurred without urban and housing policies to enable the legal settlement of the migrant population. São Paulo's urbanization can be traced back to the 1920s when the growth of the coffee plantations in the interior of the state led to a population increase in the city. This trend continued in the 1930s. The São Paulo state economy was directly affected by the stock market crash of 1929 because it was dependent on coffee grain exports. Due to the collapse in coffee prices, the country's economic model became unsustainable.

However, some of the industries already established in São Paulo had been operating below capacity. The new economic situation stimulated these industries to operate at their full capacity and, after a while, to expand. From the 1930s until the beginning of the 1980s São Paulo's industrial park flourished. These industries needed an increase in cheap labour and workers were obtained by the internal migration that peaked during the 1960s. Industrialization was supported by state policy and the internal migration was therefore important to both the economic planning of São Paulo and the industrial companies. The problem was that no housing policy was developed for these migrant workers.

The migrants could afford small lots in the urban periphery where there was no infrastructure and often no title deeds. These lots were usually purchased through small monthly payments. Over time, houses were built but the occupants could not receive title deeds to the properties because the lots were irregular. The lack of urbanization policies caused the expansion of the city towards the rural areas and divided it into two parts, only one of which was regular and had infrastructure.

This pattern started to change in the late 1970s. Due to economic factors and Federal Law No 6766/79, which regulated urban land subdivision, the offer of urban lots at affordable prices decreased. This led the poor to settle in favelas, or slums.

The 1988 constitution gave the union the power and the duty to legislate issues concerning civil law, civil procedure, urban policies and national regulations (article 22). In order to comply with the social function of the city (articles 30 and 182), the states must establish regional regulations for urban policies within their territories, while municipalities must legislate on, and control, the settlements within their territories. All levels of government are responsible for promoting housing and sanitation programmes. The constitution also declares that in order for the city to achieve its social function, property use must comply with the regulations of the city's master plan. If tenure security for the urban poor is to be achieved, there will have to be a change in the laissez-faire political practice that allowed the chaotic urban growth to unfold in exchange for votes.

THE *USUCAPIÃO* RIGHT AND 'ADVERSE POSSESSION'

Usucapião is defined as the right of tenure acquired by the possession of property, without any opposition, during a period established by law. Adverse possession is defined in *Black's Law Dictionary* as a:

> *method of acquisition of title to real property for a statutory period and under certain conditions ... there must be proof of non-permitted use that is actual, open, notorious, exclusive and adverse to the statutory prescribed period* (Black, 1990).

Adverse possession is an important instrument for securing tenure in Brazil. It originated in colonial times (1500–1822) because the Portuguese crown granted occupancy titles to its nobility, with no defined description. Today a lot of land in Brazil is without title because of confusing property registration.

In order to acquire the right to adverse possession, the applicant has to be aware that he or she is occupying somebody else's property. In contrast, the *usucapião* right can be recognized even if the occupant has a title that proves the regular acquisition of the property, but due to the absence of any formal requirement it cannot be registered. Another important difference is that adverse possession admits the acquisition of public property; *usucapião* cannot be used for claims against public property, only for private property. The *usucapião* right is discussed in this chapter.

In Brazil, property is either private or public. According to the civil code, private property does not belong to the federal, state or municipal governments (article 65). The civil code does not impose restrictions on the use and sale of private property. Property rights are conveyed by registration. Land possession is regarded by the Brazilian legal system as an extension of property rights. This kind of judicial approach leads to problems when trying to secure tenure for informal settlements. Informal dwellers can only obtain tenure after gaining a registered title deed.

Usucapião is one of the ways used to acquire property (civil code, article 530, III). There are three types of *usucapião* in the legal system: ordinary, extraordinary and special. These relate to the different statutory periods and other obligations that must be fulfilled and proved in court before the right to tenure can be declared. Extraordinary and ordinary *usucapião* are from the civil code (articles 550, 551), whereas the special *usucapião* is a constitutional right and a five-year period has been established as the statutory period of possession. The other prescribed conditions, according to the federal constitution, are distinguished by the location of the property, whether urban (article 183) or rural (article 191).

The *usucapião* definition is misleading because the right to tenure gained by it is not automatic. The possession of property without opposition during the statutory time gives a person the right to claim the recognition of tenure before the courts.

In the Federal Republic of Brazil the municipalities have budgetary and administrative autonomy. Municipalities do not have their own judicial powers and can only legislate on local matters. Claims regarding municipalities are resolved in state courts. Claims that concern the union are judged in federal courts.

Only the union has the power to legislate about civil procedure and civil rights regarding the use of *usucapião* or property rights. The municipalities legislate about urban norms in their territories but are not able to change the concepts of *usucapião* or property rights. The municipalities have the power to enforce urban property to accomplish its social function but have very limited tools available. The urban *usucapião* created by the constitution in 1988 is an instrument for urban land regularization and for enforcing the social function of urban property. To accomplish this task the municipalities have to create an administrative structure to enable the population to promote *usucapião* cases.

The urban *usucapião* is a singular type of *usucapião*. Discussions to include a special *usucapião* for the urban poor began during the 1934 national constitutional assembly. The process of urbanization had started and the judicial norms given by the civil code led to tremendous losses by buyers. Unfortunately it was not approved. A few years later the central government published Decree No 58/37 that successfully diminished the economic losses experienced by the migrant population because of the lack of methods to secure tenure. The decree also determined that municipalities must approve urban land development and that the approved parcelling plans be registered at the public registration office. These aspects were seldom respected. As a consequence, the lots bought by the migrants were parcelled without regard for the existing law. The decree lost its validity on the 20 December 1979 due to the publication of Federal Law No 6766/79.

Lots were even purchased from people who had no property titles. When the legal owner claimed the real estate property from the occupants, the occupants were left with two options: to repurchase the lot or to leave. The irregular settlements led to some social conflicts and many expulsions. At the beginning of the 1970s, during the military dictatorship (which lasted from 1964 to 1988), legislation was proposed in the national congress that aimed at the reduction of the statutory time for extraordinary and ordinary *usucapião*. The proposal was never approved.

It was only during the last national constitutional assembly (1987–1988) that, through a proposed popular amendment, the urban *usucapião* appeared. The assembly approved the urban *usucapião* (federal constitution, article 183) for the acquisition of tenure by urban settlements with the following criteria:

- five years' statutory time;
- possession must be continuous and without contestation;
- maximum lot size of 250 square metres;
- habitation only;
- claimant can be male or female, regardless of marital status, and cannot own any other real estate; and
- the right can be recognized only once in a claimant's lifetime.

The constitution did not establish rules about court procedures needed to recognize the urban *usucapião*. For this reason the right to urban *usucapião* can be recognized – as any other *usucapião* – through regulations established by the civil procedure code.

A claimant who pleads for recognition of tenure for any property on private land in any part of Brazil has to have tenure declared by the end of the procedure described by articles 941 to 945 (civil procedure code). A claimant who has possessed the property for the statutory time period of five years, for example, enters a plea for the declaration of his or her tenure based on article 183 of the federal constitution. The case has to be presented before whichever court has jurisdiction over the location of the property.

The first plea petition has to be based on a property survey. The claimant has to present, through his or her lawyer, the citation of the neighbours and the owner, and notification must be given to the federal, state and municipal treasury departments. Since the *usucapião* recognition plea is declared erga-omnes (against everyone), a legal notice must be published in the media at least three times and one copy needs to be posted in the chamber of the courts to inform those who might have an interest in the case. Finally, the claimant must request the summons of a representative of the public ministry (this Brazilian institution has responsibilities similar to that of a state prosecutor in the United States legal system, but its obligations are far wider) who must state his or her point of view about every act during the procedure.

The claimant has to prove his or her uncontested possession during the statutory time. Proof of lack of opposition is a certificate issued by the courts stating that there have been no cases issued against the claimant for the past 10 or 20 years. The law does not specify what proof must be produced to demonstrate actual possession during the legally required length of time. Although the claimant may simply offer witnesses, it is wiser to support the plea with documents that prove his or her possession. The documents can be contracts, bills or letters posted to the claimant's address during the statutory period.

The São Paulo court procedure takes two years before a final decision on tenure is declared. Tenure will be granted if the owner and neighbours are easily found, and if there is no contestation.

Brazilian courts have three levels. At the first level a judge examines the claim. At the second level – known as the 'tribunal' – the claim is examined by at least three judges. The third level is composed of the superior courts in Brasilia (Brazil's capital since 1960), which have national jurisdiction. That means that if there is a contestation, a decision on the *usucapião* claim may take an unpredictable length of time.

SOME URBAN *USUCAPIÃO* CASES IN THE CITY OF SÃO PAULO

Due to the nature of the urbanization process, there is tremendous potential for applying urban *usucapião* to secure tenure in the City of São Paulo, particularly because it is the only existing legal instrument that can be used to secure tenure on private land possessed without title. The nature of the problems that residents of informal settlements encounter in the *usucapião* procedures depends on their location in São Paulo's eastern or southern region.

The city's eastern zone was first occupied by a Jesuit mission station known as 'Aldeamento de São Miguel e Guarulhos', which served the native Indians. By the year 1710 this mission no longer existed and the region was taken over by the Portuguese colonists. The native Indians who lived there had either escaped, been killed or were enslaved. The Presidential Act (no 9.760/46) established that the area within the defunct mission territory had not been legally acquired before publication of the Act and hence it became federal property (article 1, h). Therefore, the federal government started administering the territory.

The Supreme Court determined that this Act was no longer effective because it was published shortly before the 1946 constitution and the new constitution did not include this region among the national assets. Therefore it was not recognized by the new constitutional order and could not be considered a national asset. Despite this decision, *usucapião* procedures to declare tenure over any land in the eastern part of the city confronted the federal claim. In October 1998, there were more than 9000 such claims awaiting decisions in the federal court according to the chief federal attorney in São Paulo.

In April 2000 the chief federal attorney in Brasilia published an order forbidding the federal attorney's contestation in São Paulo. Because of the protracted court procedures and never-ending lists of appeals, claims were taking up to 12 years to process and, with a bit of luck, it could take eight years to obtain a title deed. Nowadays, new *usucapião* claims do not face this kind of contestation. Unfortunately, claims admitted before the above order are still awaiting formal recognition of the union to the non-existence of its property rights. It is going to take a long time for each of the 9000 *usucapião* claims to be recognized.

A different property issue faces the settlements in the city's southern region. The principal reservoirs in the São Paulo metropolitan region – Guarapiranga and Billings – are located in the southern region. In 1979 a state law established an environmental preservation belt around the reservoirs and forbade occupation of this land. There were already about 100,000 people settled in the area when the law was passed. The occupation by additional informal settlements did not recede and today at least 1 million people are settled there. These settlements do not have sewerage systems and the reservoirs are being polluted.

Instead of establishing a metropolitan policy to solve the sewerage problem, in 1997 a new state law ordered the removal of those informal settlements built after the 1979 law. The eviction policy is not accompanied by a housing policy and so the cycle is likely to continue; the evicted people will build new informal settlements and probably be expelled once more.

A non-governmental organization (NGO) was created in 1995 to help *usucapião* in São Paulo's urban periphery. This NGO operated in 11 different settlements in the eastern and southern peripheries. It handled over 300 urban *usucapião* cases in São Paulo's court of justice before its conclusion in 1998. The cost for filing a claim to seek tenure through an urban *usucapião* claim was estimated at US$1080, including US$360 for attorney's fees.

Vila Primeiro de Outubro and Parque Amazonas

During August and September 1997 a survey sponsored by the Fundação de Amparo à Pesquisa do Estado de São Paulo (FAPESP) was undertaken by this NGO on two informal settlements. The two communities, Vila Primeiro de Outubro and Parque Amazonas, have different land occupation histories. Nevertheless, *usucapião* programmes were used to secure tenure because both settlements were on private land.

The survey aimed to demonstrate the potential efficiency of using urban *usucapião* as a mass urban regularization policy. Each settlement leader received 350 questionnaires and the following information was gleaned.

Vila Primeiro de Outubro is located in Guaianazes, in the city's eastern region. This informal settlement began with an invasion of a vast unoccupied portion of land on 1 October 1981. This property had been subdivided by its actual owner in the 1920s when it was still a rural area. Streets were created and some lots were sold to small farmers. Occupation of the area was not reported until 1981. More than 1000 families now live there informally.

After 1981 the settlement was connected to the city's water and electrical systems. But the settlement was never linked to the city's sewerage system, nor were streets and sidewalks paved. The houses that were once built with scrap wood and cardboard are now made from brick and concrete. The informal occupation left no communal space for a public square or other facilities.

Mostly migrant women answered the questionnaires. Their median monthly income was US$78, while the median monthly family income of 60 per cent of the people interviewed was US$394 (40 per cent of those interviewed did not reveal their family income). Only ten occupants were interested in joining the NGO programme. Nine filed claims but three withdrew after their claims were presented in court, for lack of funds. The federal government claimed an interest in the remaining six claims. The federal claims were rejected in the lower federal court. The appeals to the federal tribunal in São Paulo were scheduled for examination in 2001, representing a three year delay for the claimants in the commencement of the *usucapião* procedure.

Parque Amazonas is located in Parelheiros in São Paulo's southern region. It originated in the early 1970s when the owners – two married couples – decided to subdivide their rural property and sell the 220 lots. This subdivision was not conducted according to the Presidential Act 58/37 in effect at that time. This meant that the purchasers of the lots could not register their contracts after final payment had been made to the original owners.

There was an invasion of the unsold lots and this was reported. Violent clashes followed between those who had purchased their lots and those who had not. Meanwhile, no space was left for public squares or any other community facilities.

Among the residents who responded to the questionnaires, 43 per cent were male and 57 per cent were female. Most of them were migrants. Their median monthly income was US$204, and the median monthly family income of 84 per cent of those interviewed was US$431. The higher income plus a stronger leadership in this smaller community (in comparison with Vila Primeiro de Outubro) meant that 29 families showed an interest in the programme. Two of these never managed to file their claims and another two had to give up after their claims were filed.

The remaining claims were filed in different years: 17 in 1996, seven in 1997 and one in 1998. They are all still awaiting decisions. The problem facing these claims was not the federal government but finding the former owners! Proof of original ownership was required in each of the 26 cases. As mentioned earlier, the owners were two couples: one couple had moved and could not be located; the other couple was divorced, the woman had died and the man had changed his address several times. The procedure to prove that the original owners have an unknown address is very time-consuming and it has taken three years for the Parque Amazonas urban *usucapião* claimants to conclude just this part of the procedure.

Some other interesting data emerged from this survey. Residents of both settlements are not only poor and experience extreme difficulty having their right to property declared, but they also have not received local government recognition of their informal settlements and they are not linked to the city's sewerage system. The neighbourhoods are violent, there is unemployment and most of the residents are illiterate (meaning they have never attended school, or they did not finish elementary school).

The local government that records the land as a rural area does not recognize Parque Amazonas. Vila Primeiro de Outubro is considered urban and, therefore, the occupants have been sued for not paying taxes. Residents of both settlements proposed the regularization of administrative procedures to the local government in around 1990. These procedures are pending.

For Parque Amazonas to be declared an urban settlement by the municipality, urban construction has to be undertaken. At Vila Primeiro de

Outubro a survey and cartographic work has to be undertaken. In both cases the municipalities claim they do not have funds either for construction or for making a new map of the area. The local government has filed the plans of the 1920 subdivision and it charges local taxes based on that old plan in which lots were far bigger than they are today. As a result, nobody pays municipal taxes at Parque Amazonas and only a quarter of those interviewed at Vila Primeiro de Outubro pay.

Conclusions

Through the use of the *usucapião* right, the constitution intended to provide a new way to secure tenure and to enforce the social function of urban property (Saule, 1997). The importance of securing tenure goes beyond achieving the social function of urban property to fighting the social and spatial apartheid faced by Brazilian cities. It thereby helps to combat urban poverty.

The Habitat Agenda enforces the need to develop sustainable human settlements around the world. It states, 'Promoting equitable, socially viable and stable human settlements is inextricably linked to eradicating poverty' (UNCHS, 1997). Only security of tenure will bring stability to human shelter; otherwise settlements will remain precarious, increasing urban poverty.

It is apparent that more than constitutional or Habitat Agenda intentions are needed to change the reality of urban land occupation and eradicate urban poverty. More than ten years after the constitution created urban *usucapião*, and more than six years after Brazil signed the Istanbul Declaration, the goals have not been met. There are many factors working against the security of tenure policy: the conception of property as an economic value; the instability of local governments' tenure security programmes; and the identification of the types of urban programmes that can successfully secure tenure for the urban poor.

Property in itself has an economic value. Brazilians regard property as a secure asset. This sentiment can be understood because property is seen as a safeguard against the constant devaluation of the Brazilian currency. This concept of security in property also goes back to a time when inflation was not yet a national problem. Before the 1824 constitution the possibility of owning private property did not exist. Until 1822 all Brazilian land was under the dominion of the Portuguese crown. Occupancy was permitted through use concessions offered to Portuguese nobles. These concessions were made to facilitate the colonial exploitation of Brazil and obtain profits at the lowest cost to the Portuguese crown (Furtado, 1975). The colonization formula enabled the creation of a powerful agricultural aristocratic class. After independence two new economic concepts entered the national debate. The first concept was influenced by the Industrial Revolution and aimed at developing cities and placing Brazil in the emerging industrialized market. The second concept sought the continuity of the agricultural economy.

Law No 601/1850 represented the consolidation of the agricultural elite because it permitted the continuity of the existing economic model and it consolidated the concession of titles issued by the Portuguese crown (Cirne,

1988). Law No 601 gave commercial value to property with titles at the same time as it punished as a crime the occupation of untitled land. It also encouraged an immigration policy of white and free workers.

During colonial times property had no commercial value – concession titles could not be sold – while imported slaves were costly. Law No 601 reversed this situation. After 1850 property could not only be bought and sold, it could also be mortgaged. This launched the Brazilian property market and property speculation began. Since then property has been seen as an economic reservoir (Rolnik, 1997).

Despite the perceived value of property, there exists a number of judicial tools to enable security of tenure, other than urban *usucapião*. Any legal instrument chosen to obtain security of tenure and the regularization of settlements needs local policy support.

It is almost impossible to regularize every informal settlement that exists in the City of São Paulo. The difficulties found by the programme mentioned above are proof of this. On the other hand, because of the nature of the Brazilian legal system, urban reform policy must be national. It is not rational for the constitution to create a legal tool for regularization that the federal government can contest without legal grounds, as is happening at Vila Primeiro de Outubro. It is incomprehensible that a social instrument such as urban *usucapião* has to follow the same procedure as any other *usucapião*.

The regularization programmes need financial support. It is not possible to conduct a land regularization programme based exclusively on the residents' financial resources. Their income is too low. Even those who can afford the *usucapião* procedure will only receive the property titles, not sewerage connections. They may have gained security of tenure but their living conditions will not automatically improve. To improve conditions in the settlements, government polices are needed.

Stability of local urban policies is far from being achieved. This requires the political will of the mayor and the commitment of the city council. As long as they are largely dependent on the political will of the mayor, urban policies will continue to change every four years. For the past 20 years we have seen two trends in urban policies in São Paulo: one trend aims at the construction of new streets, tunnels and viaducts; the other emphasizes social programmes and the improvement of the informal city. For the past 20 years the first trend has been dominant because it is easier to maintain, as it does not demand much work from the municipal staff.

The municipality of São Paulo does not offer staff incentives to its employees. Salaries are not increased with staff productivity and staff tend to resist any extra work. Urban policies that emphasize social improvement require extra work. São Paulo needs to change its bureaucratic structure to reduce employees' resistance to work.

São Paulo is not an easy city to administer. It has a population of over 10 million people in an area of over 150 square kilometres. The city is the centre of a metropolitan region with more than 15 million inhabitants in 790 square kilometres. Its location within a metropolitan region complicates the relationship between the city administration and the São Paulo state bureaucracy.

The city administration is partially decentralized into regional departments. Department managers are selected on the basis of political criteria. This system is disastrous as the political appointees do not make good managers, and city government has been the centre of endless corruption. The first step towards improving informal settlements has to be the reorganization of the entire administrative structure in order to build permanent programmes. Specific services should be created parallel to the bureaucratic reconstruction. A special service should be created to handle urban *usucapião* cases in the city's informal settlements. Both the city and the state of São Paulo already have reasonable legal staff, who could be partially redirected to work on cases to secure tenure for informal residents. These changes require a strong political will, but it is possible.

Although ten years ago popular movements had the strength to win the recognition of the right to urban *usucapião* and have it approved by the national constitution assembly, it is hard to envisage the same mobilization to fight for a different procedure to make that right effective today. The endemic unemployment and decreasing power of the nation-state that accompany globalization have led to the weakening of popular movements. Historically, popular movements in Brazil struggled against low salaries and made claims for better working conditions. Workers' unions today are concerned with massive layoffs. The fight today is to protect jobs. The pressure for urban reform policies will not come from the informal dwellers. This pressure requires politicians who are aware of and dedicated to making the 1996 Habitat Agenda a reality in São Paulo and Brazil.

References and Bibliography

Barrufini, J T (1997) *Usucapião Constitucional Urbano e Rural*, Editora Atlas, São Paulo, Brazil

Black, H C (1990) *Blacks's Law Dictionary*, 6th edn, West Publishing Company, St Paul, Minnesota

Ceneviva, L L and Silva, H B (1993) *La Regularisation des Etablissements Irregúliers dans le Ville des Pays en Developpements,* Coordenação de Alain Durand-Lasserve, Workshop in Mexico sponsored by the World Bank

Cirne Lima, R (1988) *Pequena História Territorial do Brasil*, ESAF, Escola de Administração Fazendária, Brasília, DF

da Silva, J A (1995a) *Direito Urbanístico Brasileiro,* Malheiros Editores, São Paulo, Brazil

da Silva, J A (1995b) *Curso de Direito Constitucional Positivo*, Malheiros Editores, São Paulo, Brazil

Dinamarco, C R (1995) *A Reforma do Código de Processo Civil*, 2ª edição, Malheiros Editores, São Paulo, Brazil

Fernandes, E (1998) *Direito Urbanístico: coordenação de Edésio Fernandes*, Editora Del Rev, Belo Horizonte, Minas Gerais, Brazil

Ferraz, T S Júnior (1987) *Introdução ao Estudo do Direito*, Editora Atlas SA, São Paulo, Brazil

Furtado, C (1975) *Formação Econômica do Brasil*, Editora Nacional, São Paulo, Brazil

Gomes, O (1980) *Direitos Reais*, Editora Forense, Rio de Janeiro
Holston, J (1993) *Legalizando o Ilegal: Propriedade e Usurpação no Brasil*, Revista Brasileira de Ciências Sociais, São Paulo, Brazil
Kelsen, H (1973) *Teoria Pura del Derecho*, Editorial Universitaria de Buenos Aires, Buenos Aires, Argentina
Linhares de Lacerda, M (1960) *Tratado das Terras do Brasil*, Editora Alba Ltd, Rio de Janeiro, Brazil
Margarido, A B (1991) 'Sobre a Aplicabilidade da Função Social da Propriedade', tese de doutorado, Faculdade de Direito da Universidade de São Paulo, São Paulo, Brazil
Markoff, J (1998) 'The Future of Globalization', Lecture paper presented at XVI International Sociology Association World Congress on *The Future of Globalization and Democracy*, Montreal, Canada, August 1998
Montesquieu, Baron de (1979) *De l'Esprit de Lois*, GF-Flammarion, Paris
Nunes, P (1954) *Do Usucapião*, Livraria Freitas Bastos SA, Rio de Janeiro, Brazil
Rodota, S (1986) *El Terrible Derecho: Estudios sobre la propriedad Privada*, Editorial Civitas SA, Madrid, Spain
Rolnik, R (1997) *A Cidade e a Lei*, FAPESP/Studio Nobel, São Paulo, Brazil
Saule J'nior, N (1997) *Novas Perspectivas do Direito Urbanístico Brasileiro*, Ordenamento Constitucional da Política Urbana, Aplicação e Eficácia do Plano Diretor, Sergio Antonio Fabris Editor, Porto Alegre, Brazil
Taschener, S P (1996) 'Compreendendo a Cidade Informal', in *Desafios da Cidade Informal*, UNCHS (United Nations Centre for Human Settlements, Quênia), Nairobi
UNCHS (1996a) *An Urbanizing World*, Global Report on Human Settlements, Oxford University Press, Oxford
UNCHS (1996b) *Report of the Recife International Meeting on Urban Poverty*, UNCHS, Quênia, Nairobi
UNCHS (1997) *The Istanbul Declaration and The Habitat Agenda*, UNCHS, Quênia, Nairobi
Werna, E (1995) 'Urban Management and Intra-urban Differentials in São Paulo', *Habitat International*, Vol 19, No 1

Chapter 8

The Right to Housing and the Prevention of Forced Evictions in Brazil

Nelson Saule Junior

The Brazilian legal system was reformed by the 1988 Brazilian constitution. For the first time, a specific chapter was dedicated to urban policy. This policy contains a set of principles, responsibilities and obligations for the government to adhere to. It sets out the legal and urban instruments to be applied in order to correct the situation of environmental degradation and social inequalities in the cities, thereby ensuring dignified living conditions for the urban poor.

Considering the increase in forced evictions, how does one solve urban environmental conflicts and guarantee compliance with the constitutional principle of the social function of property? This needs to be addressed by the judiciary and other institutions that administer justice, as well as by federal, state and municipal government agencies responsible for environmental protection. A set of measures must be adopted to change Brazilian urban legislation in order to create laws and legal instruments aimed at promoting social and territorial inclusion and integration in the cities. The structure of government must be democratized in order to guarantee popular participation in the management of the city, as well as the establishment of new public spheres to resolve collective conflicts, as in the case of forced evictions. This will ensure that the urban rights of the population living in the city, especially the marginalized groups, are truly respected and protected.

THE RIGHT TO HOUSING AND THE ROLE OF THE STATE

The materialization of the right to housing, reaffirmed as a human right at the second United Nations Conference on Human Settlements (Habitat II, Istanbul), requires that nations guarantee this right through the promotion of urban and housing policies.

The right to housing is also recognized as a human right in various international human rights declarations and treaties of which Brazil is signatory, including: the Universal Declaration of Human Rights of 1948 (article XXV, item 1); the International Pact for Economic, Social and Cultural Rights of 1966 (article 11); the International Agreement for the Elimination of all Forms of Racial Discrimination of 1965 (article V); the Agreement for the Elimination of all Forms of Discrimination against Women of 1979 (article 14.2, item h); the Agreement on the Rights of Children of 1989 (article 21, item 1); and the Human Settlements Declaration of Vancouver of 1976 (Section III (8)).

The basis of this right to housing is incorporated and integrated in Brazilian law and legal procedure in the 1988 Brazilian constitution (article 5, paragraph 2), which determines that,

> *The rights and guarantees expressed in this constitution do not exclude others that stem from its regime and from the principles that it adopts, or from international treaties in which the Federal Republic of Brazil is part.*

This means that the list of rights expressly foreseen in the constitutional text is illustrative and not binding because, based on the regime and principles adopted in the constitution, new rights can be established. It also means that the rights recognized and protected by the international treaties, of which Brazil is a signatory, are incorporated into Brazilian law, as is the case with the right to housing. Based on this constitutional norm, the right to housing becomes one of the fundamental rights of the Brazilian legal system.

For the right to housing to have legal and social effectiveness, positive action is required by the state via the execution of public policies, especially the promotion of an urban and housing policy. The housing policy must establish agencies, legislation, programmes, plans and tools of action to guarantee this right for all of its citizens. It is important to recognize that this requirement does not mean that the government is obliged to supply a residence, or home, to each citizen. In reality, this requirement has two aspects. The first is of an immediate nature, the prevention of any regression in housing law. It includes prohibiting measures and actions that make it difficult or impossible to exercise the right to housing; for example, the prohibition of a housing system and policy that includes exclusionary and discriminatory measures that impede access to housing rights for a large portion of the population. This, unfortunately, has been the role of the housing finance system, which makes the reformulation of this system obligatory.

The other aspect is the promotion and protection of the right to housing by government, which must intervene and regulate private sector activities related to housing policy. This includes: regulating the use of and access to real estate property, particularly in urban areas, in order to meet its social function; regulation of the land market; the establishment of systems of social interest housing finance; and the regulation and establishment of guidelines concerning urban land use, the right to build, tax policy, rental policy and usage concessions for housing purposes.

Meeting these requirements makes effective the right to housing. In particular, the following measures are necessary in order to exercise this right: the adoption of financial, legal and administrative instruments for the promotion of a housing policy; the formation of a decentralized, democratic, national housing system, with mechanisms for popular participation; revision of legislation and instruments in order to eliminate norms that carry any type of restriction and discrimination in the exercising of rights to housing; and the distribution of resources that promote a social interest housing policy.

Housing policy: Obligations and responsibilities of the state

According to the constitution (article 21, insert XX), the federal government has the exclusive responsibility to establish guidelines for urban development, including housing, basic sanitation and urban transport. Upon establishing housing policy guidelines, the federal government must, for example, define the objective criteria for the application and destination of resources for the housing finance system. Included must be criteria for the use of these resources by the states, municipalities, private agents and social agents, in social interest housing programmes and projects.

It is the sole responsibility of the federal government to legislate on civil law (article 22, insert I) and to organize the private relationships concerning the right to housing and property law, in particular, a system of property rental, ownership, use and benefit, and the tools of property transfer, such as contracts for purchase and sale.

The federal government is also responsible for establishing general guidelines for land policy. Brazil has a serious problem with land access, which is demonstrated by various conflicts over ownership in urban and rural areas. These conflicts result in physical violence and occasionally death. Because of the large number of properties not utilized, and therefore without a social function, the civil code must be reformed, as well as civil procedure concerning the mechanism of possession and ownership. Special forums to judge these conflicts must be established to facilitate obligatory procedures, public negotiating and conciliation hearings, with the participation of public agencies. A judicial inspection, that is the physical presence of a judge at the site of litigation, must issue a judicial decision in order to improve the process of the hearings and to substitute the instruments of injunctive measures that have generated more social conflicts than solutions.

The federal government must establish general norms of urban law via the federal urban development law. According to federal urban development law (article 4, paragraph 4), it is necessary for the regulation of the urban mechanisms for compulsory construction or land division, progressive taxes on urban land and buildings, and misappropriation for the purposes of urban reform. The municipalities, to guarantee compliance with the social function of property based on the master plan, must apply these.

The responsibility of the states under complementary legislation ensures the establishment of both state housing law and urban law, in order to apply these policies in an integrated manner by the municipalities. The states are responsible for instituting a system of state policy with its own agencies and instruments. These policies must be destined particularly for the metropolitan areas.

Concerning housing policy (article 23, insert IX), the federal government, the states and the municipal governments must promote housing construction programmes, improvements of housing conditions and basic sanitation. The social interest housing programmes ensure that those excluded from the housing market are served, for example, by the regularization of land and the provision of infrastructure in favelas.

Since the municipality is ultimately responsible for the execution of urban policy, it must develop a local housing policy. In order to confront and address the needs of the housing problem, the municipality can establish municipal government housing agencies and institutions. It can implement a municipal housing system that includes a system of housing policy management with popular participation, for example, through the establishment of councils, financial mechanisms, social interest housing programmes and urban tools.

To develop a housing policy the municipality must institute a master plan in order to establish guidelines and instruments for the use and occupation of urban land, methods of cooperation between the private and public sector, and criteria for the social use of urban property. To regularize land use, the municipality can also adopt specific laws concerning social interest housing and urbanization plans for precarious settlements, operations of social interest, transfer of the right to build, a right to create space, special social interest zones and concession of real use rights.

TYPES OF URBAN CONFLICTS

The Social Policy Institute and the University of São Paulo conducted a survey of cases of urban environmental conflict in the city of São Paulo (10 million residents). These conflicts were the object of judicial actions and decisions by the state judiciary of São Paulo from 1996 to 1998. Based on the survey, the following types of conflicts relating to forced evictions were identified:

- individual and collective occupations of public lands (favelas);
- individual and collective occupations of private lands by social groups;
- collective occupations of public buildings in the central region of the city by groups and movements of *cortiço* residents;
- individual and collective occupations of vacant spaces below bridges and overpasses;
- establishment of irregular or clandestine popular subdivisions in areas of water source protection by companies and real estate developers, community associations and housing cooperatives;

- irregular establishment of housing projects and urban subdivisions by the government; and
- irregular establishment of housing projects by community associations through community work projects.

These types of conflicts were also identified in the municipalities that compose the São Paulo metropolitan region, known as ABCD, which is the centre of the automotive industry.

The increase in informal settlements in the central regions of the city of São Paulo, identified by the survey, reveals the actual growth of the real city. This increase has led to a situation of environmental degradation and to the social and territorial inequality of Brazilian cities. Two cities exist within the same territory, the 'legal' city where the privileged live and the informal but 'real' city where the majority of the population live in conditions of exclusion and marginalization.

There are many factors to be considered regarding forced evictions that occur in the 'illegal' or informal cities. The poverty and destitution of the majority of the urban population, which in recent years was aggravated by the growth in urban unemployment, impedes access to housing and the formal property market. The lack of positive steps by government to address the situation has created an increase in social and territorial inequality. This is characterized by the lack of urban and housing policies with programmes and public and private resources to guarantee the low-income population access to dignified housing. Government is responsible for the failure to inspect the use and occupation of urban land and for not enforcing legal mechanisms to combat real estate speculation in the cities, thereby ensuring the social function of urban property.

The urban legal order excludes the territories in the city occupied by the poor and marginalized urban population. It is formed from a set of legislation that contains elitist urban standards for land division, land use and occupation, and construction of urban land. This legislation serves the needs of that portion of the population with an income. Historically these standards have been defined by the interests of the property market, which determines certain areas of the city to have an elevated or reasonable standard of urban environmental quality of life (through the supply of infrastructure, equipment and services). This situation generates the appreciation of property in these areas due to public investments, as well as protecting the interests of the population in the neighbourhoods with a reasonable or elevated urban environmental quality of life. It also generates a set of laws concerning land use and occupation and division, with discriminatory rules that create social and territorial inequality. The established urban norms do not recognize settlements as legal settlements where the poor population and vulnerable groups live.

There is no direct relationship between the existence of urban regulation and urban precariousness. The municipality of São Paulo, for example, has a high degree of urban regulations: a master plan; laws for land use and occupation; a construction code; land division laws; laws for social interest housing, urban operations and coordinated operations; and laws for transfer of right to build. On the other hand, there is a high degree of urban precariousness, considering that more than 1 million people live in favelas, 3 million live in collective housing or *cortiços* and 2.5 million people live in irregular or clandestine subdivisions in the peripheral neighbourhoods that offer precarious conditions of infrastructure, equipment and urban services. The urban legal order in Brazil must be modified to change the position of government agencies and the judiciary, which give preference to rights concerning private property over those pertaining to the public sphere of urban property.

In Brazilian legislation, the norms of the civil code and the code of civil process that regulate and protect the individual right to property are not in keeping with the constitutional principle of the social right to housing. The legislation protects owners of urban land and property and this is detrimental to the public and social interest of the right to housing, cultural patrimony and the environment. This strong protection is found in rental relations, issues of possession use and benefit from urban property. The changes in the urban legal structure should be directed at the establishment of urban laws that contain legal mechanisms. Local governments could then use these laws against the owners of land and buildings who do not comply with the social function of the city, especially urban speculators. The measures include a progressive property tax, mandatory building and subdivision regulations, and the appropriation of property for the purpose of urban reform. This would democratize access to decent housing for all who live in the city. The changes in the urban legal order also imply the establishment of laws to recognize the legality of informal but consolidated settlements. This would:

- impede the constitutional violation of human rights through forced evictions of the population living in informal settlements considered illegal in the field of urban legislation;
- recognize the right of all those who live in informal but consolidated settlements to have formal title to the urban area in which they live, to obtain decent housing with urban infrastructure and the supply of public services (basic sanitation, transport, public lighting, health care and education); and
- create obligations for government to improve urban and environmental conditions through urbanization plans aimed at the establishment of urban infrastructure and the supply of public services.

The majority of the population who live in informal urban settlements have great difficulty protecting and promoting the defence of their rights, often because they lack knowledge about their rights as citizens. In the case of forced evictions, the affected population rarely has legal assistance to promote the defence of their rights. This is despite the fact that the Brazilian constitution determines that such assistance should be supplied by the state without cost to people in need. The institution responsible for offering this service is the public defender, which should be established by the federal government and the states. Since few states have created public defender offices, this service is very precarious in Brazil. The existing legal assistance agencies are insufficient to meet the needs of the poor. One of the ways to overcome this deficiency is for municipalities to create legal assistance services to offer legal orientation and promote the protection of the rights of socially needy groups. The municipalities can also establish partnerships with non-governmental human rights organizations to conduct the service.

The organizational structure of the Brazilian judiciary does not contain judicial forums with the capacity and efficiency to deal with collective conflicts; for example, urban environmental conflicts that result in forced evictions. In the lower courts, cases that result in forced evictions are usually judged in a civil court where the judge normally deals with individual interests and with the compliance or non-compliance with private obligations. This situation is also found in the state courts. The Brazilian constitution allows for the creation of special courts for the solution of urban conflicts that cause large social impact to the city, but this has not yet been undertaken by any Brazilian state.

Another problem with the judicial system is that highly populated municipalities and metropolitan regions are not permitted to establish municipal courts to resolve issues of urban interest. Only the federal government and the states can have their own courts. In many cases of forced evictions, the local government has been responsible for promoting a mediating and negotiating process between the population affected by the request for removal, the government agencies, the judiciary, those responsible for the forced eviction request in court and other social actors involved and interested in resolution of the conflict. In many of these cases the intervention of local government has been much more efficient than that of the judiciary to establish accords between the parties in order to avoid forced eviction. These experiences reinforce the proposal for the judiciary to create specialized public forums within cities to resolve urban environmental problems, which would make the participation of local government in the mediation and negotiating process mandatory.

ACTIONS TAKEN BY HOUSING MOVEMENTS

In Brazil there are two important popular movements to protect and defend the right to housing. One of these organizations is the National Struggle for Housing Movement. To open negotiating space with the Brazilian government and obtain support from society, it has among its actions the collective occupation of vacant land on 3 June of each year in important Brazilian cities. In order to criticize forced evictions publicly, this movement annually compiles a report on the number of evictions and families who were victims of violations of the right to housing. The report is sent to Brazilian government authorities so that measures can be taken to resolve the reported conflicts.

The other organization is the National Union of Housing Movement. This movement's strategy is to promote marches and long-distance caravans calling for the creation of popular, self-managed housing programmes (under the community work system), where the management of resources and of the housing project is the responsibility of the community organization (cooperatives or not-for-profit civic associations). The proposal is to create these housing programmes with public resources from federal, state and local governments.

In the city of São Paulo this movement has various housing organizations among *cortiço* residents. *Cortiços* are the most precarious and inhuman housing alternative for the poor. They are usually old properties that were divided into small rooms, often housing more than five people, including children, adults and the elderly. They also often present fire and explosion hazards because of poor wiring and the concentration of stoves and gas tanks. The small number of bathrooms and the unsafe hygiene conditions make these areas a public health risk. In addition to the precarious housing situation, their legal situation is also insecure. There is a lack of formal rental relationships between the residents and the owners. This is due to the lack of a contract, the lack of proof of payment of rent and because the intermediary who administers the *cortiço* charges rents incompatible with family incomes and overcharges for water and electricity. This situation is aggravated by the connection of many of these intermediaries with the police and criminal organizations. It is extremely common for occupants of the *cortiço* to be evicted by court order because of the lack of documents that prove they are tenants and not illegal occupants of the property, or prove compliance with rental payment obligations.

To change this extremely inhuman situation the *cortiço* movement has in the past two years occupied vacant government buildings in downtown São Paulo to demand a policy for the *cortiços*. Their main demands are:

- the undertaking of housing projects in vacant public buildings to benefit the *cortiço* population;
- the revitalization of abandoned areas downtown with housing projects for the *cortiço* population; and
- the creation of a rights protection programme for this population, seeking to avoid the continuation of forced evictions and economic and moral exploitation of the people who live in the *cortiços*, as well as legal assistance services and special inspection units to deal with violations of the right to housing.

MEASURES TO COMBAT FORCED EVICTIONS

There is a need for a federal urban development law that aims to regulate the urban policy chapter of the Brazilian constitution. The urban legal instruments need to be regulated by the following federal laws: compulsory land division or construction; progressive urban property tax implemented over time; misappropriation for the purpose of urban reform; the onerous concession of the right to build; urban requisition; right to pre-emption; surface rights; squatters' rights; and the concession of constitutional use. It is also the responsibility of the federal government to establish general rights about land policy. There is a need to revise municipal urban legislation, especially the laws concerning urban land division and the use and occupation of urban land. Social and territorial integration needs to be achieved to reduce social inequality in the cities.

In order to meet the principle of equality, urban legislation must be reviewed to establish norms for social inclusion, through the recognition of the urban rights of the population who live in the real or informal city. These include the right to housing, urban infrastructure, public equipment such as schools, day-care, community and sporting centres, and to urban services (transport, basic sanitation, waste collection and treatment, electricity, telecommunications and postal service).

The recognition of urban rights in the field of urban legality is fundamental for the establishment of government obligations to implement social policies, for the allocation of resources for the execution of programmes, and for plans and projects which seek to implant urban infrastructure and equipment, as well as to offer urban services in territories of the informal cities where the environmental urban quality of life is highly precarious.

The review of urban legislation has the goal of constituting a set of norms and urban tools that promote the social and territorial integration of the population who live in conditions of poverty, by recognizing the urban rights of this population and the fair distribution of the wealth of the city (property and tax income).

The master plan is the basic tool of development policy and of urban expansion of a municipality, through which are established the basic demands to order the city according to the Brazilian constitution (article 182). The master plan should be instituted by municipal law, and is obligatory for municipalities of more than 120,000 inhabitants. Article 182 also establishes that compulsory subdivision or construction is applied by the municipal government to ensure that owners of urban property subdivide, build or utilize this property in instances when the property is considered as not utilized or under-utilized in terms of the master plan.

'Urban operations' is a tool established by municipal law that includes an integrated set of interventions and measures coordinated by the municipal government with the participation of owners, residents, users and private developers. It is implemented in urban areas that need to be recovered and revitalized.

Regulatory measures concerning property use and ownership

Concession use is a measure applied by the government to transfer the 'right to use' to families who are settled in public areas, for residential purposes, without the transfer of property. It offers security to collective occupations and impedes forced evictions. The concession use aims at regulating the legal situation of people who live in public areas such as favelas. It is a mechanism that guarantees the social purpose of a public area, thereby avoiding property speculation, as the urban area is not privatized.

Urban squatting is a measure established in the Brazilian constitution (article 183). It recognizes the right of special urban *usucapião*, a form of adverse possession in private (never public) urban areas of land up to 250 square metres. This right must be declared by the courts, which must issue an order to register title over the squatted area. However, only a few municipalities in the metropolitan region of São Paulo are adopting programmes and measures that allow urban squatting to be solicited in court.

Municipal legislation establishes the criteria, requirements and urban standards for the production of social interest housing and the urbanization and recovery of urban areas considered to be in the social interest for housing purposes.

THE CREATION OF SPECIAL ZONES OF SOCIAL INTEREST

Special zones of social interest can be created in urban areas occupied by favelas, *cortiços* in collective residences, popular subdivisions, and vacant and underutilized urban areas. These zones have been the most efficient tools adopted to avoid forced evictions. Where there is a high number of land ownership conflicts or there is housing that will result in forced evictions of needy social groups, municipal law can define these areas as special social interest zones. This will legally guarantee the permanence of social groups in the area of conflict through the legal recognition that these areas are destined for the housing of the social groups that live in these informal settlements.

Based on this tool, the judiciary can judge the requests for eviction and removal of families who occupy public or private areas in favour of the social groups, and establish a process of negotiation between the owner of the area, the residents and the government. This regulates the legal situation of the population and promotes urbanization and improvement of the urban conditions of these areas through urbanization programmes. Some Brazilian municipalities have utilized this land regulation tool to combat forced evictions.

There is a need for municipalities to establish a system of democratic management of the city, through the establishment of public, municipal, sectoral and regional forums for popular participation. These should have decision-making power over the use of public resources, the implementation of public policies and power to mediate and negotiate urban environmental conflicts. The democratic management of the city is a political system that meets the constitutional maxim of direct democracy (article 1, sole paragraph of the Federal Constitution), which seeks to assure participation of the population, especially of the marginal population, in the decision-making process concerning important issues in the city. The right to participate in decision-making allows the creation of institutional channels that can serve as spaces to mediate and negotiate urban environmental conflict between the social sectors involved.

The formation of special forums for mediation and conciliation of urban environmental conflicts is necessary because often the decisions issued by the judiciary aggravate the situation of social and territorial inequality. These decisions simply determine the dismantling of informal but consolidated settlements, or the clearance of favelas that are already consolidated but located in a public area formally declared a 'green area', which in reality was always an area destined for housing purposes by marginalized social groups. It is important to re-establish the judicial function of promoting social justice, which requires taking the initiative in conflict resolution through mediation, discussion and negotiation.

The special forums to judge these conflicts should have procedures that establish the requirement for public hearings and for negotiation and conciliation, with the mandatory participation of government (federal, state and municipal agencies). Mandatory judicial inspection should also be established. This implies the physical presence of judges and prosecutors at the place of the legal dispute as a requirement for the issuing of a judicial decision, and to substitute the mechanism of injunctions, which have served more to aggravate than to resolve social conflict.

Special forums for urban environmental conflicts should be established in metropolitan areas, especially in municipalities with a population above 1 million and in municipalities with a high level of urban environmental conflicts.

There is a need for the legal actors (judges, prosecutors, public defenders, attorney generals, lawyers and legal assistants) to be trained to promote social justice in the solution of urban environmental conflicts and in the field of human rights, especially in the international human rights protection systems (including treaties and conventions and international human rights agencies). The norms of the international and national systems of human rights protection should be adopted and applied, as well as the constitutional principles and objectives of Brazil which include: citizenship and human dignity; direct democracy; the eradication of poverty and marginalization; the reduction of social inequality; the principle of equality; the principles of the social functions of the city and property; and environmental protection. These should be used as criteria for the solution of urban environmental conflicts.

The way forward

At the beginning of the 1990s a proposal to establish a national housing council and a national housing fund with popular participation was presented to the national congress by popular initiatives. The proposal was signed by 1 million voters and aimed at establishing a housing policy for the low-income population.

The proposed law being considered in Congress seeks, as a substitute for the special housing commission, to establish a national social interest housing system. The goals of this system are to organize, make compatible, monitor and support the activities of agencies and entities that act in the social interest housing field, and to make available and promote access to urban and rural housing for the low-income population by implementing a subsidy policy.

Social interest housing is defined as that destined to serve the low-income population. Beneficiaries will be those with a monthly family income below US$15. They cannot be property owners, prospective buyers or have a concession of rights to use any other residential property.

The system will be composed of the following: the national housing council; the ministry responsible for the housing sector (currently the urban development secretary); federal, state and municipal public administrative agencies; state and municipal housing councils; and regional and metropolitan entities. The system also includes the housing agencies or equivalent entities and companies, cooperatives, consortia, community associations, foundations or other private associations of a profit or not-for-profit nature, which perform activities in the field of social interest housing, as well as the financial institutions that operate in this sector. The government-managed savings bank (Caixa Econômica Federal) is the operating agent for the system's resources.

The proposed system has a hierarchical structure, with the national housing council as the central entity. The composition of the council will include representatives from the federal government union centres and legally established popular housing organizations. It will be responsible for: approving the national housing policy; establishing guidelines, strategies and tools and the priorities of this policy; approving the programmes for allocating resources of the national housing fund; establishing the subsidy policy for the system; approving the annual and multi-year programmes presented by the state housing secretariat and other responsible agencies of the system for the utilization of the fund resources; and regulating and investing other resources destined for housing received by the Brazilian savings and loan system (SBPE).

The national housing fund is planned as an instrument of this system. It will centralize federal resources destined for the social interest housing programme and for financing the activities and programmes developed within the system. The fund's resources would be broadly constituted as follows: 70 per cent from the resources of the guaranteed fund for time employed (FGTS); 20 per cent from the resources received in savings deposits by the entities that constitute the SBPE; and 10 per cent from the net income of lotteries, betting services and other federal contests of any nature, and federal budgets destined for this purpose.

Among the programmes for which the resources of the national housing fund will be used are:

- housing construction, infrastructure (sanitation and urban facilities inherent in housing);
- production of serviced lots, provision of infrastructure for favelas and improvement of housing units; and
- intervention in *cortiços* (tenement type housing) and in collective rental housing, acquisition of construction material, production and acquisition of property for social rentals.

Concerning the distribution of resources by the national housing council, at least 60 per cent will be distributed to attending families with monthly incomes of less than six minimum wages, and a maximum of 10 per cent will be destined for deserving families with monthly incomes between 10 and 15 minimum wages.

The criteria adopted in the promotion of the subsidy policy should include the following: the concession of subsidies to ensure housing to applicants with family incomes up to four minimum wages; 80 per cent of the minimum value of the monthly payment as the maximum limit of the subsidy; and an inverse relation between family income and the subsidy.

Guidelines concerning the integration of state and municipal entities call for the formation of national housing secretariats and housing councils, or equivalent organs, the composition of which must consider the participation of the public and private entities directly linked to the housing field, in addition to community representatives, especially popular housing organizations. For the same reasons, states and municipalities should establish special funds for the implementation of the social interest housing programmes, in order to allocate financial resources captured at the state or municipal level and to complement those issued by the national housing fund.

The project requires a decentralized and democratic system for the implementation of a socially oriented housing policy. This will be achieved by including as criteria for the state and municipalities their constitution of funds, specific budget allocations and housing councils with popular participation. In order to use the funding resources, the state and municipalities will have to develop social interest housing programmes in which the agents are popular housing cooperatives or unions.

It is extremely important that the national housing council is given the responsibility to regulate the application of the resources of the SBPE. This will allow the resources of savings legally allocated to the low-income population to have a destination based on the national housing policy guidelines, removing from the central bank the responsibility of defining the investment of these resources.

Chapter 9

Favela Bairro: A Brief Institutional Analysis of the Programme and its Land Aspects

Sonia Rabello de Castro

The Favela Bairro programme, meaning from slum to neighbourhood, has received national and international recognition because it is an example of efficient state intervention in urban housing through government investment. The programme aims to integrate slums into the legal city by providing access for the population of the favelas to basic services and infrastructure, as well as the regularization of its public and private spaces.

It is important to note that this programme was instituted on the basis of a local government policy decision taken in 1994, and was carried out by technical staff from the municipal government. This unusual institutional arrangement was possible because the federal system of government in Brazil differs from other models like those found in the USA or Germany. In Brazil municipalities are autonomous entities with their powers prescribed by the federal constitution, and not by a delegation of state entities, as usually occurs in other federations.

These conditions allow municipalities sufficient constitutional autonomy to decide and implement their own public policies, without interference from federal or state government. However, they have to do so with their own funds and assume responsibility for the associated risks. Thus in the case of Favela Bairro, the government of the City of Rio de Janeiro assumed all responsibility for the conception and implementation of the programme. The federal government must, however, approve international financing of parts of the programme, as determined under Brazilian law.

The implementation of the Favela Bairro programme was possible because of the political determination on the part of the government of the City of Rio de Janeiro. Brazilian municipalities, besides being autonomous federated entities, also hold a specific set of powers to provide all – or almost all – public services that were necessary for the development of the programme.

It must be emphasized that the federal model of the Brazilian state enabled – through a simple political decision by the head of the local executive branch and the consequent approval of the budget by the local city council – the implementation of the programme. The programme required of the municipality of Rio de Janeiro an executive, administrative and financial capacity that is found in very few Brazilian municipalities.

The Favela Bairro programme targeted areas designated as favelas, or slums. Favelas are largely residential areas where poor populations took possession of vacant public or private areas of land. Their existence in Rio de Janeiro dates back to the 19th and early 20th centuries. Curiously, the name favela comes from Favela Hill (Morro da Favela), a hill located in the central area of the city. Since then the name has been used for informal settlements. Informal settlements in the 19th century existed on the slopes of the Castelo and Santo Antônio hills, but they were demolished in favour of urban projects aimed at modernizing the downtown area of the city during the early decades of the 20th century.

The reasons for occupation of city land have remained the same. A population lacking the resources to secure housing seeks a place to live in vacant urban areas, or other areas not being used by their owners (public or private), which are located close to their places of work or accessible to transport.

> *It is worth noting that the occupation of these areas has a special character, that is, the search for a place to live... They are occupied by people attracted to the combination of modern transportation and available space* (Porto Rocha, 1999).

The lack of sufficient income, which made it impossible for the very poor to rent or buy standard housing (under the norms of municipal law), was worsened by various factors. During the first quarter of the 20th century, public authorities implemented urbanization projects in Brazil. They cleared a number of areas where poor residents used to live in collective houses close to their places of work. There was a complete absence of government policy to supply new homes for the displaced population. Municipal urban legislation not only ignored but also aggravated the housing problem by forbidding traditional communal houses in the new urban areas.

In some ways, all of these aspects of urban public policy in Rio de Janeiro remained the same throughout the 20th century. This allowed favelas to become the most common and systematic social reality in the city, because they solve the housing problem of the low-income population. The only difference is that formerly they were located in central parts of the city, which became the subject of real estate speculation because of urbanization. These areas were highly valued because of urban public investments and especially because of their prime business location. Today favelas are precarious settlements located in peripheral areas and locations close to places of work, mainly on hillsides in the city, or other high-risk areas.

Favelas are different from other kinds of irregular urban land occupations, called irregular or clandestine developments. What differentiates the other developments from favelas is the issue of ownership. Irregular and clandestine developments contradict the rules of public authority because the owner of the plot does not follow the approved project, carries it out irregularly, or does not submit the project for approval by the local urban authority. Nevertheless, in all cases, the owner sells the plot, but without proper documents.

These private sales have a certain validity because there is proof that there was an intention to dispose of the asset, the buyer operated in good faith and legal deeds exist that, according to civil law, provide the essential elements necessary for a legal transaction between private individuals. These documents, although they have a limited and sometimes debatable legal value, are the basis for the regularization of ownership and lend legitimacy to the transaction.

It is important to stress that Brazilian society, in general, has little or no information about the illegality of acts that are committed when urban rules relating to the use of private land are not followed. Most of the time, people ignore the rules and, consequently, also the illegal act. Nevertheless, for the majority of the population, especially the poorest, the general consensus is that the owner holds command over his or her property, even if the transaction is not officially recognized. Thus, if he or she sells it, the transaction is valid, no matter how it is undertaken. People are therefore not too concerned about buying irregular or clandestine lots.

The sense of security experienced by the 'new owner' has important consequences: he or she will be inclined to respect the limits of 'his or her property'; more money will be invested in permanent construction; and services such as water, sewerage, transport and health clinics will be requested from the local authorities.

From the outset favelas are precarious and irregular. This form of occupation is essential to the survival of many people. Occupation is continuous and people settle within the limits of the local physical conditions. The area is settled without criteria for the definition of the occupied area, or for the number and size of the lots. Unplanned accesses and passageways are built among the dwellings in a haphazard fashion. No claims are made for basic services because this 'luxury' is not a priority for those who are only concerned about survival and a roof over their heads.

The inhabitants of favelas acknowledge the precariousness of such occupations because nobody has given them any guarantee of ownership. For many years they will face great uncertainty about whether they will be allowed to remain. As a result, no investments are made in improvements to homes, common spaces or services. Likewise, there is little incentive to create a system for receiving newcomers to the settlement. The fact that favelas are spaces occupied without the consent of the owner creates an enduring situation of uncertainty and insecurity – the stigma of precariousness.

The state has remained dissociated from the favelas. Favelas are like an urban limbo in comparison with formal and regular society. The laws of the state are not applicable, nor do its police or courts enforce them. As with any social group, and for their own survival, residents create their own internal rules. Self-appointed leaders and chiefs enforce coercive measures to ensure that their 'laws' are respected. Public services do not reach them: there is no public transport, often because there is little access for motor vehicles; there is no sanitary infrastructure, such as water and sewer systems; and the electricity supply is weak, especially public lighting. Only some favelas, usually the oldest ones, have elementary schools, and these are in very poor condition. Very few have health clinics and there is no waste collection or, if it exists, it is inefficient. As a result, environmental and sanitary conditions are extremely poor.

There is no systematic urban planning, that is, the adaptation of legislation for land use (for division, building and utilization of the land), since there is no urban legislation applicable to these areas. In the master plan of the City of Rio de Janeiro, favelas are shown as blank areas, in other words, lacking any legal recognition whatsoever. As already noted, land tenure (when it is not violent and/or clandestine, which would make it illegal under private law) is uncertain and it is very difficult to demarcate areas due to irregular housing arrangements. Given such a multiplicity of uncertainties, the struggle for power is inevitable, as is social violence and criminality, which operate beyond the limits of state control.

All these aspects make favelas antithetical to the standards that regulate the formal city. Favelas are chaotic spaces compared with the notion of official urban regularity. Nevertheless, after a century of existence, the favelas are home to almost 1 million people, 20 per cent of the urban population of the city.

The Favela Bairro programme

The Favela Bairro programme was introduced in 1993, when the newly elected government of the City of Rio de Janeiro took office. At that time local government did not include a housing secretariat in its administrative structure. The question of public intervention in the favelas was considered to be the jurisdiction of the urban planning secretariat. Prior to 1993 local government had implemented a few sporadic programmes for reforestation, basic water and sewerage, in the form of *mutirão* (collective community work). But there had been no systematic public intervention in the favelas.

The Favela Bairro programme arose from the acceptance of the existence of favelas as an urban social reality and the urgent need for the official system to adapt to this reality in the best way possible. The name of the programme represents the goal to change favelas into neighbourhoods in order to integrate them into the city. It is hoped that with the implementation of the programme, favelas might even lose the name 'favela', and be called by their true names like other recognized neighbourhoods: Borel, Mangueira, São Carlos, Copacabana, Tijuca, Ipanema, etc.

In the past favelas were basically residential, but today, from the perspective of the city as a whole, they are more complex and problematic. The housing question is the original and principal purpose of the programme, but it also encompasses all the basic and fundamental aspects of urban administration, with the implementation of services and infrastructure. For this reason the master plan of the city (article 139) established that the responsibility for city housing policies would be that of the agency in charge of urban planning.

Thus, the secretariat of urban planning was responsible for the favelas until March 1994, when the municipal government granted the new housing secretariat total jurisdiction over the implementation of housing policies in the City of Rio de Janeiro. Two important points arise in this regard: the housing policy of the city is not limited to the Favela Bairro programme; and, although the programme is conceptually considered an aspect of housing policy, it is basically a very complex urban planning programme encompassing social goals and a plan for the regularization of land ownership.

In summary, the Favela Bairro programme is essentially an urban planning programme in the broadest sense. From the point of view of its goals, Favela Bairro is a housing programme, since it aims at the legal and urban regularization of predominantly residential areas.

Despite the conception of Favela Bairro as urban planning, due to a political decision and for purposes of administrative convenience, it was decided that the housing secretariat would administer the programme.

The general scope of the Favela Bairro programme and the housing policies associated with its implementation have been included since 1992 in the law of the master plan, article 138, and in the city charter that sets out the organization of the municipality of Rio de Janeiro.

> *Urban development policies will obey the following precepts: urbanization, land regulation and registration of the areas occupied by the* favelas *and low-income population without the removal of residents unless required because of physical limitations that put its inhabitants at risk. In this case, the following rules are to be followed:*
>
> - *A technical report is to be issued by the responsible agency.*
> - *The interested community has to participate in the analysis and definition of solutions.*
> - *If displacement is necessary, the population has to be settled in areas close to their original dwellings, or close to work (Article 429).*

As the executor of the programme, the municipal housing secretariat directs and coordinates the work of a special task force formed of managers from the highest levels of local government, named the 'executive group for popular settlements' (GEAP). The priority of this group is the interaction and coordination of programmes carried out by each one of the municipal secretariats responsible for the implementation of specific aspects of the Favela Bairro programme.

In its effort to transform the favelas into regular neighbourhoods, the programme seeks to provide them with the entire infrastructure and all the services typical of neighbourhoods. For this reason, in each favela there are many different infrastructure projects: those that provide accessibility by main and secondary roads; provision of water and sewerage; the availability of electricity; public areas for leisure and sports; public works to contain erosion; as well as environmental projects, such as reforestation. Other important aspects are the removal of houses from high-risk areas and the clearing of places where projected passageways, roads and common social spaces are to be built.

Except for the replacement of a displaced house, there is no provision for new homes in the programme, but it is probable that improvements brought about by infrastructure projects will motivate individuals to invest more in their houses.

Beyond the implementation of urban planning projects, it is important to note that the Favela Bairro programme envisions the introduction of important social development programmes. These include sports, leisure and culture, health care and job training and educational programmes, especially those implemented through the municipal network of primary schools. The implementation of these diverse social projects makes the programme unique and gives it real possibilities of sustainability and future development by the communities themselves.

It is important to stress that in implementing these projects through local government, the existence of favelas as an urban social reality has been acknowledged. Local government has had to invest in their improvement, as it has always done for the other neighbourhoods in the city.

Once again, we return to the fact that, because the Brazilian federal model grants autonomy and legal authority to municipalities, allowing them to develop their own programmes, local government can carry out all the projects of the programme without state or federal legal authorization. Municipal government has exclusive jurisdiction over urban planning in the city and thus has the power to decide on zoning, building codes and other parameters for the use of land in any part of the city. This allows for local government to consider various urban models and their relevance for the areas occupied by the favelas, accepting them as they are, with only slight adjustments.

The municipality also has jurisdiction to install basic infrastructure and services, such as water, sewerage (there are legal questions about the capacity of the municipality to provide water and sewer systems, related to the eventual ability of state government to provide these services), electricity and areas for leisure; waste collection, public roads and passageways; and the prevention of erosion on hillsides and areas of risk.

Finally, the municipality may implement, by its own means, all social projects that affect the community, especially those related to public health, education, environment, labour, social development and culture.

According to the Brazilian constitution, all three levels of the government, federal, state and municipal, have executive authority to implement housing projects. Article 23 of the federal constitution states explicitly, 'It is the common responsibility of the union, states, federal district, and municipalities ... to promote housing programmes and the improvement of housing conditions and basic sanitary programmes...'.

By this constitutional measure, Brazilian legislation included the promotion of housing within its activities for the public good. Housing is basically a private economic activity subject to the private investment market. From the constitutional perspective, the economic activity of the real estate market for housing is not restricted to private investors.

Consequently, this authority given to the Brazilian municipalities implies responsibility for its administration. At the same time, it allows local political authorities to approve the development of projects as comprehensive and complex as Favela Bairro by a relatively simple political decision, independent of the approval of authorities from state and federal government.

LAND TENURE ISSUES OF THE FAVELA BAIRRO PROGRAMME AND ITS REGULARIZATION

Favelas are a form of occupation for which the Brazilian legal system does not yet have mechanisms in place to protect the inhabitants, despite the fact that they may be poor, and acting in good faith, occupying unused or abandoned land.

In the City of Rio de Janeiro, until the municipal charter was passed, every occupation that had become a favela was under constant threat of removal. Consequently, for decades, due to an urban culture developed over the course of the 20th century, these urban residential centres were excluded from public and social investments. Furthermore, no effort was made for their integration into a system of legal ownership or safe tenure.

Another serious consequence was the exclusion of the favelas from normal public services, especially their marginalization from the system of public security, and from legal protection. These settlements developed despite the state and thus there was little knowledge or application of official laws. The favelas developed their own code of rules and 'rights of occupation'. Today these rules are 'enforced' by local groups who often use weapons and violence to maintain their power. Linked to the criminal world, they are the 'owners' of the favelas.

Owners neglect the land where the favelas are located for two reasons. First, the state is often unaware of its own holdings and, in any case, it does not react to unauthorized occupations by rich or poor even if it has knowledge of them. Second, in cases of privately held land, the areas are considered unattractive by their owners because they are located in rather inaccessible places. Favelas are often on the slopes of mountains or close to lakes, areas still excluded from the speculative property market, which invested in centrally located, flat areas (from which poor people were displaced at the beginning of the 20th century) and, more recently, in the coastal area.

As a result of property market speculation, which occurred during the 1960s and 1970s, favelas close to Lagoa Rodrigo de Freitas and the elegant quarter of Gávea, close to an expensive private university, were removed. Nevertheless, most of the favelas on the slopes of the hillsides along the coastal area of the city, such as Copacabana, Leblon and São Conrado, still remain. In fact, the largest favela in Rio de Janeiro, 'Rocinha', is located in São Conrado, a wealthy neighbourhood of Rio.

The land question is the most basic problem for the social and urban integration of the favelas. Hence the Favela Bairro programme emphasizes regularization as one aspect of the programme. Land regularization is explicitly mentioned in all proposals relating to the programme and yet it is the only aspect that has not shown any progress. In sporadic cases municipal authorities provided title deeds to lots, but this occurred only under special conditions: for example, in public areas that were not hilly and thus with a type of occupation that facilitated delimitation of the lots. Nevertheless, even in these cases, the land regularization remains unfinished, as only title deeds have been issued to the tenants without any follow-up measures. Legal and social effects related to the stability of property holdings have not been monitored.

The regularization programme has not been carried out according to the programme proposals because of the Brazilian legal system, which involves complex legal, urban and civil law questions. In this regard, even the political decision from local government and the resources to support the decision do not suffice.

The formulation of a specific programme of action is needed, one that takes into account factors beyond the jurisdiction of local government. These include: civil legislation, included in the federal power that defines the terms of the rights to property; the legal nature and effects of property rights in general, contracts and condominium agreements; location and general rules for the division of land, family and succession rights; and the rules of procedural law (ways to access judicial power). Federal law also defines the rules that determine acquisition by the effects of tenure (*usucapião*), the rules of public registry and expropriation.

The regularization of the favelas cannot be considered simply in terms of the implementation of urban projects. A specific and complex plan must be developed. Many studies will be necessary to consider all the legal variables presented by each situation. These studies will have to clearly outline alternatives, include them among other legal possibilities and adjust everything in a frame of predetermined objectives to be implemented under different circumstances.

Besides these legal questions that extend beyond the legal power of the municipality (questions predetermined by the federal legal system), land regularization will be subject to two other partners: the judiciary power and the affected population.

In Brazil conflicts of interest and questions concerning rights to rural or urban land are part of social and legal history. It is the judicial power that, in the end, interprets the rule of law over conflicts. Without the participation of the judiciary it would be impossible to build legal theories or even enact laws because of the possibility that, in the medium term, they could be judged incompatible with constitutional law. It is the decisions of the judiciary that are the safest source of a new interpretation of the right to property.

For this reason, the participation of the judiciary power in the conception and implementation of the regularization of land programme is vital. This is neither a simple nor a small task, because it is an act that involves the participation of another constitutionally independent state branch. Furthermore, legal authorities tend not to get involved in social problems, intending, by this stance, to maintain their 'neutrality'.

The involvement of the interested communities in the programme is not any easier. It is believed that the regularization of land will only be effective if the community is the primary agent responsible for its implementation. Obviously this is a long process that will initially require government support. It is the community, however, that will need to know their rights and pursue ways to achieve them. This is the practice of citizenship, which exists only when the individual consciously and effectively seeks it.

Although the task appears complex, it is not impossible. It will initially require a clear understanding of the overall problem and of the objectives to be achieved, of the legal instruments available to be used, and of the ways the project can be carried out, the agents involved, objectives and time limitations. Because of all those factors, the land regularization programme will require considerable determination and patience because it can only be implemented and consolidated over the long term.

Specific problems

From a legal perspective there are many problem areas around land regularization. Two aspects that are fundamental to the project are the design and form of occupation of the favela and the question of rights with respect to laws of registration.

The occupation of the favelas was disorganized because of the absence of tenure security. As a result, the form of occupation and style in which houses were built is completely incompatible with other traditional forms of urban occupation that are determined by urban legislation and civil law. Houses are built one on top of another to maximize the available space. Often it is almost impossible to determine where the boundaries of a house start or finish. It is even less possible to determine the boundaries of the lots in which these houses are located. It is consequently very difficult to provide title deeds for these properties.

It is not viable to regularize favelas using an urban plan to demarcate individual lots, even though this is the legal requirement for their registration at the registry of deeds. This is the reason for proposing a programme that is not in accordance with legally approved civil and urban standards. Instead of adapting the favelas to the legislation, the legislation must be adapted to the reality of the favelas in order that they may be integrated into the city as officially recognized neighbourhoods.

There is a need to find an innovative solution to the problem of how to update the title deeds of houses. The forms of construction and the occupation of the favelas are totally incompatible with the timing, the format and costs set out by the registry of deeds. The Favela Bairro programme has included plans for the demarcation of public areas, main and secondary roads, space for infrastructure, and areas and buildings to be used by the community. However, if the demarcation of public areas represents an important advance because it allows the clear identification of those areas and leads the population to respect their boundaries, there are also similar challenges associated with their title deeds. Up to now it was not possible to register them at the registry of deeds because a detailed plan of the whole area was required, as well as their original registration.

With regard to the regularization of the unofficial systems in the favelas, there do not appear to be any local norms or conventions that specify alternative building models in the favelas where the programme has been implemented. In this instance there is also an enormous difficulty in adapting traditional approaches to alternative forms of construction and occupation, which are so different from the favelas, and in finding consensus.

The acquisition of title deeds, as is required by Brazilian legislation, is difficult to adjust to a land regularization project for the favelas, not only because of questions around urban planning but also because of questions relating exclusively to property. The registration of an urban property, to obtain the title deeds, includes the assumption of facts that are very difficult to find or maintain in areas like the favelas. These facts are related to the conditions surrounding the achievement of title deeds and the location of the plot.

In many instances ownership is not clearly established and in other cases the question of ownership involves conflict. There are also instances where plots are under legal litigation and others, not so rare, where the plots do not officially exist as they have never been registered. The latter situation, which seems simple, in actual fact complicates the programme of regularization. In Brazil, in principle, there is 'no land without ownership'; it is the registration that has the effect of creating the title deed of ownership.

Nothing is registered at the registry of deeds unless it is precise, official and legally defined in the legislation. The title deed is only issued according to precise formal definitions. The claimants must have their personal situations identified properly, a condition which is very seldom found in the favelas. However, because of its complexity and inaccessibility, the registry of deeds has always been the most commonly used instrument and the easiest form of real estate 'grilagem' (the means by which individuals who are not owners become, by illegal or irregular means, the owners of land occupied by other tenants without a title deed) in Brazil.

Conclusions

The Favela Bairro programme is undoubtedly a project that has been successfully implemented, in terms of its urban planning objectives. The favelas represent a possible and already consolidated housing option for a significant number of people in Rio de Janeiro. The programme therefore addresses the housing policies of the municipal government.

The implementation of the Favela Bairro programme was only possible because it was conceived from a simple political decision. According to the Brazilian constitution, local authorities have jurisdiction over the execution of social projects and public services. The programme involves complex implementation because it deals with the realization of different type of projects and services such as urban planning, projects for urban and environmental infrastructure, and social and land regularization projects.

The land and housing question forms the basis for the irregularity and precariousness of the favelas. The uncertainty of the residents' legal situation results in their social exclusion, marginalization from access to benefits and services, and urban and social rights overall. The exclusion is also evident in the lack of protection by police forces and legal guardianship from the state. The completely marginal condition of the favelas in the face of the rule of law causes the communities to make their own 'rights', which are enforced by groups and local 'authorities' who are beyond the control of the state.

The section of the programme related to land regularization is still to be developed. Land regularization does not enjoy the same conditions and facilitating factors as the other projects in the programme because there are questions and aspects that are not the exclusive decision of local authorities.

The land regularization project of the Favela Bairro programme, because of its complexity, requires a specific plan of action in order to be viable. The following objectives need to be clearly established: ownership or assurance of ownership; the issue of title deeds in the short or medium term; the legal means to implement the project; the participation and involvement of local agents; the knowledge of related problems; and the solution of critical social, urban, economic and bureaucratic issues, as well as possible legal alternatives.

If the land regularization project is not implemented soon, the whole Favela Bairro programme will be affected because of the unsolved question of the precariousness of ownership, which generated the irregularities and the total exclusion of the favelas in the first place.

References

Porto Rocha, O (1999) *A Era das Demolições,* Secretaria Municipal de Cultura, Coleção Biblioteca Carioca, Rio de Janeiro, p75

Part 3 South Africa

Chapter 10

Security of Urban Tenure in South Africa: Overview of Policy and Practice

Lauren Royston

SOUTH AFRICA'S LONG HISTORY OF TENURE INSECURITY

In South Africa the questions of tenure security and urban access have historically been intricately linked. A policy on native locations and limitations on access to residential rights was firmly established as early as 1910. By the late 1950s and 1960s urban population resettlement was a cornerstone of the apartheid strategy to reduce the size of the urban African population. South Africa has a long history of territorial segregation and tenure insecurity. The implementation of the apartheid vision saw urban inhabitants 'superfluous' to white needs resettled in ethnically defined 'homelands' (or reserves or bantustans), the development of industries on the borders of bantustans and the denial of urban residential rights. This period saw the introduction of the infamous Group Areas Act and it was from this time that millions of people were evicted from their homes within the urban areas. Freehold settlements such as Sophiatown, Lady Selbourne, Marabastad, District Six and Cato Manor were destroyed to force black residents to move to formal townships on the periphery of the 'white' cities and towns. Thousands of black people were evicted from urban areas altogether and removed to bantustans, small areas of mainly rural land set aside for black occupation.

Mass public housing construction occurred in black townships (black group areas in the former Republic of South Africa territory) on an unprecedented scale at this time. Some 500,000 dwellings were financed by government and built by the larger municipalities (Arrigone, 1994). The tenure rights of occupants were restricted to ensure temporary residence. A black person, qualifying for the infamous 'section 10' rights, could obtain a permit, issued by the municipal authority, to occupy a house in a black urban area. Only certain people were eligible for these rights in terms of which a black person 'qualified' to be in an urban area. Until 1978, there were three major forms of tenure in black urban areas, called section 6, 7 and 8 permits. They amounted to little more than permission to occupy land, allocated mainly to a male household head, without the possibility of encumbering, subletting, selling or bequeathing the property (Emdon, 1993).

Permanency was only envisaged within the homelands, where tenure forms mostly took the form of communal occupation under the commonage system, or under the aegis of traditional leaders, although the land remained vested in the South African state. Although freehold rights were conferred in some cases in the homelands, most of the land is still registered in the name of the state.

Despite attempts to the contrary, the urban black population displayed a considerable propensity to grow, and informal settlements expanded enormously in the late 1960s and have continued to grow to the present. In the bantustans unregulated settlement growth was rapid, especially in those areas located close to towns and cities.

By the 1970s the growth of informally settled areas was extensive. In the bantustans they far overshadowed any formal housing that the new administrations were able to deliver. Illegal shacks built in the backyards of formal houses within existing black townships were the main form of homelessness in some parts of the country in the 1970s, until the advent of free-standing squatter settlements in the 1980s (Mashabela, 1990). Resistance movements began to make demands for housing and other social goods a focal point for popular mobilization (Baumann, 1998). Population management was collapsing, influx control was being ignored and informality was being extended. Long before formal changes in policy, the cornerstones of the apartheid regime had begun to crumble with the de facto breakdown of influx control and unplanned urbanization (Royston, 1999).

The location and quantity of land set aside for black South African occupants was tightly controlled through legislation such as the Black Communities Development Act and the Group Areas Act (Royston, 1998a). Where land was released by provincial authorities to serve the settlement needs of migrants and new generations of black urban dwellers, remote location and inadequate service levels led to conditions of extreme overcrowding in formal housing stock (Royston, 1998a). Rapid migration in the 1980s, linked to economic pressures in the homelands and the abolition of influx control, led to settlement close to the white urban periphery, in the black urban townships, in overcrowded public accommodation, backyard shacks and informal settlements. Community struggles around housing were an integral part of the broader urban struggle against apartheid, and land invasions and boycotts against rent, service charges and later bonds were strategies adopted by civic organizations in the ungovernability campaign intended to bring down apartheid (Royston, 1998b).

Land invasion on the periphery of established group areas emerged as an alternative strategy for ensuring land access. The condemnation of informal settlers and squatters by the state and by formally settled communities ensued (Kok and Gelderblom, 1994). The location of informal settlements – mainly on the far peripheries of the metropolitan areas – was influenced by the relative strength of the core city and weaknesses of peripheral administrative structures such as the black local authorities and tribal authorities in some areas (Hindson and McCarthy, 1995).

In the face of a widening political outcry around the housing question and the evidence of acute need for delivery, a more conciliatory approach developed towards informal settlements. By the mid-1980s black South Africans were accepted as permanent residents for the first time, with the introduction of 'orderly urbanization'. Government needs to address stability in the context of growing popular opposition found expression in the housing sector in privatization, a series of changes in tenure-related law and the introduction of site-and-service schemes. This coincided with private sector interests in a secure market and financial investment in black urban areas.

In 1978 it had became possible for 99-year leasehold rights to be held in black townships (Emdon, 1993). The passage of the Black Communities Development Act in 1984 created a statutory right to 99-year leaseholds. It became possible to register these rights in 1986 and the holders of regulation 6, 7 or 8 permits could obtain these rights. Black urban dwellers could now purchase new houses built by the private sector. In addition, full freehold ownership rights for black urban dwellers were created by the Act in 1986 for the first time. Previous forms of tenure could be converted into ownership, provided that a township register was opened, entailing a survey and surveyor general approval of the plan, prior to opening the register in the deeds office. As a result of demands made by civic associations, a process of privatizing state rental stock at no charge to the occupants began in earnest, called the 'transfer of houses' process.

At this time government intervention in the form of the provision of a limited capital subsidy, coupled with private sector delivery of site-and-service schemes, began. This trend reflected a worldwide tendency toward harnessing the successes of informal housing and increasing abandonment by governments of public housing programmes (Urban Foundation, 1991). It found expression in the Independent Development Trust (IDT) Policy capital subsidy for serviced sites with registered individual ownership, the de facto housing policy of the apartheid state in its dying years. This scheme provided over 110,000 serviced sites throughout the country, including such places as Orange Farm, Khayelitsha and Inanda (Baumann, 1998). The required provision of individual ownership was intended as an instrument of securing tenure and as an asset against which finance could be raised. Payment of the subsidy to the developer was linked to the finalization of transfer, ensuring developer responsibility for the completion of construction, the hand-over of the services to the local authority, township establishment processes and registration of titles (Robinson et al, 1994). The capital subsidy scheme was heavily criticized for its minimal package: a serviced site and tenure without top structure, the slow rate of consolidation, the poor location of sites and the lack of flexibility of the policy approach (Urban Foundation, 1991). The site-and-service approach became a target of mobilization against the South African state and contributed to galvanizing the urban struggles of civic associations, led by the South African National Civic

Box 10.1 From Permits to Individual Ownership: The 'Transfer of Houses' Process in Gauteng Province

In 1990 the Greater Soweto Accord was concluded between the 'committee of ten' (which represented the Soweto Civic Association), the municipality and the provincial administration, signalling, among other things, an agreement to 'transfer houses to the people'. The Accord linked ownership by residents to the termination of the service charges and rents boycott. Negotiations, which began in the metropolitan chamber in 1991, were concluded in 1994 and agreement was reached on a process for transferring rental houses to the people. The negotiated principles were approved as the norm in the provinces. In the course of 1993 the central government announced a discount benefit scheme, in terms of which the tenants of state-financed houses could purchase them at historic cost, less a discount of R7500.

The enabling legislation for the transfer process is the conversion of certain rights into the Leasehold Act, promulgated in 1988. This law required that provincial administrations determine ownership of the subservient rights, which had been granted in black townships, administer a transparent legal process and register title into leasehold. Conversion into full ownership was made possible when the Act was amended.

In Gauteng rented houses are being transferred in former black townships, former 'own affairs' areas (former white, Coloured and Indian areas) and serviced sites in less formal townships. A 16-step process has been developed which begins with the invitation for claims and ends with the registration of transfer (Stevens, 1999). Area-based institutions, called housing transfer bureaux, have been set up within each municipal area as a joint venture between local and provincial government. Tenants and any other persons who think that they should be regarded as the rightful tenants are invited to submit claims. Claims are submitted at the housing transfer bureau office and the tenant data are verified against available records. In the event that the claimant is not the registered tenant according to available records, or that sufficient data are not available, an enquiry is undertaken. The investigation into current property information occurs at the bureau offices. Where tenancy cannot be verified, disputed cases are adjudicated and family disputes are mediated. Experts appointed from a panel (approved by the local authority) adjudicate disputes. The sale and transfer are then administered in line with current procedures.

The major challenge confronted by the transfer of houses case has been to devise and implement a transparent and legitimate process for beneficiary identification and registration of tenure (Stevens, 1999). Determining who should receive the benefit is complicated by uncertainty surrounding who the rightful tenant is due to: incomplete or incorrect records of tenancy; changes due to death, divorce etc which may not have been recorded; innumerable informal changes in tenancy; and a variety of occupancy patterns resulting from overcrowding, including the occupation by more than one family of a single unit (Royston, 1998a). The title documentation requirements have been found to be highly technical and poorly understood, leading to the suggestion that many of these need not be included. In less formal settlements experience has demonstrated the need for a more proactive approach than a purely claims-led process, which can be lengthy (Stevens, 1999).

Organization (SANCO). At the same time political violence in the hostels was reaching its peak, giving impetus to the establishment of the national housing forum (NHF) in 1992, of which SANCO, the African National Congress (ANC), unions and business interests were a part. The inclusiveness of the housing policy negotiation process is the subject of much debate, with detractors arguing that landless and community interests were absent (Baumann, 1998). The NHF evolved into a negotiating forum on housing policy, with the result that new housing policy was complete by the time of the first elections (Narsoo, 2000). The housing white paper was published in 1994.

Historically tenure insecurity has been used as an instrument of urban exclusion. Black South Africans were denied ownership rights until the recent past when the dynamics of both social change and policy failings resulted in adaptations. By the advent of democracy in 1994 the primary urban tenure challenges lay in protecting people against eviction, securing tenure in former homeland towns and informal settlements, and addressing tenure in public rental stock (already well on the way, through privatization, to individual ownership) and informal tenure in backyard shacks and for subtenants and sharers.

Main forms of informal settlement

By 1994 South Africans were holding land under a variety of tenure systems, including freehold and customary tenure and various forms of tenancy. While racially based laws have been removed, many people still face tenure insecurity. According to the department of national housing, approximately 58 per cent of all households (4.8 million households) have secure tenure (ownership, leasehold or formal rental contracts). About 9 per cent of all households (780,000 households) live under traditional, informal, inferior and/or officially unrecognized forms of tenure, primarily in rural areas. An estimated 18 per cent of all households (1.5 million households or 7.4 million people) live in squatter conditions, backyard shacks or in overcrowded conditions in existing formal housing in urban areas, with no formal tenure right over their accommodation.

As well as providing insight into the nature and extent of tenure insecurity, these statistics reflect how settlements which are insecure in tenure terms are categorized in the housing white paper. There are squatters, backyard shack dwellers, sharers and subletters and an undifferentiated category of insecure rural households. No reference is made to the irregular subdivision category that is so prevalent in India and Brazil. This could be attributable to the enormous research gaps on informal settlement, observed by Huchzermeyer (2001a), with informal settlement research exhibiting a bias towards invasions that took place in former white group areas. While references are made in South African literature and policy discourse to informal practices of 'allocation', the term 'subdivision' is seldom applied in informal contexts, being reserved for the formal and technical activity undertaken by professionals. Informal allocation tends to apply especially to the loaded 'shack lord' term applicable to processes of land transaction characterized by violent conflict; the process of 'shack farming' (a form of irregular subdivision characterized by rental); and the land allocation role played by traditional authorities. These settlement processes, which are forms of irregular subdivision, are viewed in particularly pejorative terms (illicit practices undertaken by unscrupulous individuals preying on the defenceless poor). The Prevention of Illegal Eviction from and Unlawful Occupation of Land Act, 19 of 1998 (see Box 10.4), outlaws the practice of individuals transacting in land which is not theirs to begin with.

Despite its use in the housing white paper, the term 'squatter' has negative connotations in South Africa, and preference is given to the use of informal settlement (Huchzermeyer, 2001b). 'Informal settlement' has become a fairly ubiquitous, but undifferentiated, term, which does not account sufficiently for the diverse nature of informal settlement processes or their physical manifestations. Processes of informal settlement include squatting (by land invasion or encroachment), informal rental (including subletting and sharing, which includes the backyard shack phenomenon and some inner-city tenure arrangements) and irregular subdivision as outlined above. Furthermore, as Huchzermeyer (1999) elaborates, Informal settlement means officially unplanned and unauthorized settlements. South African literature often does not draw a clear line between 'informal settlement' and 'informal housing', the latter including shacks on serviced sites, in back yards of formal township houses, and on invaded land. This has led to much confusion in the interpretation of research.

Backyard shacks (informal rental or subtenancy arrangements on sites mainly within former African townships) or overcrowding within existing township housing stock (relations of subletting and sharing) are both important forms of informal access to shelter for the poor in South Africa. A survey in 1993 in six formal black urban townships estimated that 40 per cent of the surveyed population lived in backyard shacks and that a further 15 per cent were tenants within the formal dwellings. Accordingly, a possible 55 per cent of the surveyed population were renters and sharers (Khan, 1995). Tenure conditions in these examples are not as precarious as in squatter situations, as some kind of consent arrangement has been made. Control over race zones had particular impact on the extent of backyard shacks and overcrowding within black areas, vis-à-vis illegal subdivision outside them. Furthermore, the significance of backyard shacks as a form of shelter is geographically varied according to proximity to a former homeland boundary, where access to land was much easier than in 'white' South Africa (Crankshaw, 1993). It is in the towns within these areas that activity in the informal allocation of land by traditional authorities is observable. The allocation was informal because traditional leaders had no legal jurisdiction to allocate land in former homeland towns; nevertheless they exacted charges of various sorts in exchange for land access and other services (Development Works, 1999c).

LEGAL AND POLICY FRAMEWORK AFFECTING URBAN TENURE SINCE 1994

Since the democratic elections in 1994, the housing delivery backlog has driven the agenda, rather than tenure regularization or security of tenure, through the national housing subsidy scheme. Underpinned by a particular understanding of supply and demand factors, expressed in the NHF, this approach promotes delivering new housing at a scale to cope with the backlog and new demand, and addressing system constraints to the rapid release of land. The latter found expression primarily in the Development Facilitation Act (DFA) and some innovative provincial programmes for land release. The approach also tends to de-emphasize informal land delivery and opposes land invasions on the basis that they are counter to the formal, programmed approach of government and tantamount to 'queue jumping'.

The institutional framework affecting tenure

In the constitutional negotiations leading up to 1994, housing was formulated as a concurrent competence between national and provincial government. This arrangement was a surprise outcome of South Africa's constitutional negotiations, as expectations were high that housing would be a national competence (Narsoo, 2000) and housing had certainly been treated that way in the housing policy negotiations in the NHF. In practice this means that national government makes policy and provinces implement it, that national government allocates subsidies and provinces approve their disbursement. Provinces, therefore, have the power to control the types of housing projects that get approved, including the kinds of developers (Baumann, 1998). Housing delivery has traditionally been the preserve of provincial government; provinces evaluate proposals, allocate subsidies, set minimum standards, monitor projects and administer the subsidy system (Engelbrecht et al, 1999). Due to municipal capacity constraints, provinces may remain key players in housing delivery, despite the objective of accrediting local authorities to undertake housing projects. The extent of the impact of developmental local government (particularly the formulation and implementation of local housing strategies and targets as part of the municipal planning process) on shaping local housing strategies and their relations with organizations of civil society in doing so, remain open questions.

The constitutional protection of property rights

Berrisford (1999) argues that property rights protection is an important legal hurdle to clear in reforming South Africa's system of urban law, especially in relation to two fundamental questions.

> *To what extent can the state diminish rights – or interests – enjoyed by landowners in its endeavours to meet broader developmental and environmental needs? ... under what circumstances (is) compensation ... payable to people*

> *whose rights or interests in their urban property are reduced by state actions?* (Berrisford, 1999, p12).

The answers lie in the way that the property clause contained in section 25 of the 1996 constitution is interpreted:

> 1 *No one may be deprived of property except in terms of law of general application, and no law may permit arbitrary deprivation of property.*
> 2 *Property may be expropriated only in terms of law of general application: for a public purpose or in the public interest; and subject to compensation ...*
> 3 *The amount of compensation and time and manner of payment must be just and equitable, reflecting an equitable balance between the public interest and the interests of those affected* (Republic of South Africa, 1996).

Berrisford (1999) suggests that there are strong indications that the right to property is no longer a super right, but simply one of many, including those to housing, land reform and a healthy environment. The notion of property protection that favours the established urban elite's interests is increasingly being eroded. For example, the Prevention of Illegal Eviction from and Unlawful Occupation of Land Act (see Box 10.4) makes substantial inroads into what was previously the private fiefdom of the owner (Berrisford, 1999). The Extension of Security of Tenure Act (see Box 10.3) illustrates how rights in land, if not land itself, can be redistributed within a constitutional framework that respects a qualified right to private property (Roux, 1999).

National policy and programmes affecting tenure

Since the first democratic elections in 1994, a host of new national laws and policies have been enacted and developed. It is within the policies emanating from the departments of housing and of land affairs that national policy guidelines for urban tenure security are to be found. The housing white paper claims that security of tenure is a cornerstone of the government's approach to providing housing to people in need (Republic of South Africa, 1994). Subsidy policy has been designed to accommodate a variety of tenure options on the basis of a limited state contribution to be driven by private (individual) investment, credit finance and where possible the sweat equity of the owner. Box 10.2 describes the housing subsidy scheme. The elevation of individual private home ownership above other forms of secure tenure is rejected by the government in the housing white paper. The principles, goals and strategies of the policy were transformed into legislation in the form of the Housing Act (107 of 1997). The nationally binding principles for land development, contained in the Development Facilitation Act (67 of 1995), state unequivocally that land development should result in security of tenure and provide for the widest possible range of tenure alternatives (Principle 3[1][k]). South Africa's land policy advocates that people should choose the tenure system which is

Box 10.2 The Housing Subsidy Scheme

The cornerstone of the housing policy is the national subsidy scheme, which offers lump sum or one-off 'capital' subsidies for lower income households earning less than R3500 per month. The amount is provided on a sliding scale in relation to income, ranging from R5500 to R16,000 (as from April 1999).

Monthly beneficiary income	*Project linked capital subsidy*	*Consolidation subsidy*	*Institutional subsidy*
up to R1500	R16,000	R8000	R16,000
R1500–R2500	R10,000	–	R16,000
R2500–R3500	R 5500	–	R16,000

There are five subsidy forms. The project-linked subsidy is applied to housing projects and provides for individual ownership. This subsidy gives individuals access to a housing subsidy to acquire ownership of, or upgrade an existing, property or to purchase/build a new property. Institutional subsidies are allocated to an institution as opposed to an individual and provide for alternative forms of tenure. Consolidation subsidies are available to individuals who have received housing assistance under the previous government in the form of a serviced site. The discount benefit scheme (being phased out) allows long-term tenants of public rental stock to receive a discount on the historic cost of a property (see Gauteng Province experience presented in Box 10.1). The hostel upgrading programme provides assistance for the upgrading of publicly owned hostels.

appropriate to their circumstances and the land reform programme aims to extend greater tenure security to South Africans under a diversity of tenure systems (Republic of South Africa, 1997). From these references, it is clear that the policy framework establishes the significance of secure tenure as well as the promotion of choice in tenure form.

Legislation influencing tenure

Since 1994 three key pieces of tenure legislation (presented in Box 10.3) have been passed which are directed at securing tenure for vulnerable occupiers. However, these laws are limited in application to areas occupied by tribes on land owned by the state (Xaba and Beukman, 1999). They do not apply to the majority of urban tenure situations of informal settlement upgrading or privately owned land. The Upgrading of Land Tenure Rights Act, 112 of 1991 (ULTRA), promulgated before 1994, was intended to upgrade informal land rights and permits to occupy to ownership. This act is similarly limited in application and suffers from lengthy procedures and an individual ownership bias.

Urban tenure reform and urban land in general is a much less politicized issue than rural land. Urban land struggles are treated as a technical, developmental matter of urban planning (Baumann, 1998). Urban tenure is most frequently addressed in the multiple planning laws for land development

Box 10.3 Post-1994 Tenure Legislation (Xaba and Beukman, 1999)

The following three laws give an indication of how tenure security is dealt with in the land reform programme. They are primarily directed at preventing eviction and give people the right to have a say in access to, and use of, the land they occupy.

Extension of Security of Tenure Act, 62 of 1998: This law provides occupiers of rural or peri-urban land (ie land that is not in a proclaimed township) with a legal right to live on and use that land and protects them against eviction.

Land Reform Labour Tenants Act, 3 of 1996: The Act provides for the protection of labour tenants against arbitrary evictions and makes provision for the acquisition of land for existing labour tenants who will be able to access a land reform grant for this purpose.

Interim Protection of Informal Land Rights Act, 31 of 1996: This law protects people who live on, use, have access to or occupy land, particularly in the former homelands, against eviction. It gives people who reside predominantly in areas under traditional leadership the right to be heard regarding the administration of the land that they occupy. The law is intended to lapse once more permanent protection is enacted into law.

or township establishment, which were closely tied to segregation and control. They also favour individual ownership and new developments (Development Works, 1999b). Before 1994 the key elements of the legislative framework for planning were: the national Physical Planning Act (of 1967 and 1991); provincial planning ordinances of the four provinces of 'white' South Africa; planning laws drawn up by former bantustan administrations; and regulations drawn up under the Black Administration Act of 1927 and the Black Communities Development Act of 1984 (and its predecessors) which governed African urban race zones (Berrisford, 1999).

The DFA, a transitional measure introduced in 1995, was constructed as a key intervention in addressing the land and housing backlog and playing 'the numbers game', which the housing policy ascribed to legislative constraints, among other things. The DFA set out to deal with the perceived legislative and procedural obstacles to the desired rate of land development by introducing new measures to facilitate and expedite projects and to bypass bottlenecks in existing regulations (Republic of South Africa, 1995). Designed as a short-term, transitional measure pending the outcome of more thorough processes to transform the country's chaotic system of planning law, the DFA was intended to facilitate the speedy delivery of land for development purposes. The system of land development introduced by the DFA provided for an alternative route for land development applications. Provincially created tribunals could override laws from the apartheid era which impeded positive land development and speed up decision-making processes, especially in relation to so-called 'reconstruction

Box 10.4 Noteworthy Post-1994 Urban Reform Legislation (Berrisford, 1999)

The following four laws deal with land occupation, land development and planning.

The Restitution of Land Rights Act, 22 of 1994: applicable in both rural and urban areas, this law addresses the restitution of land rights lost by any South African as the result of a racially discriminatory law passed since 1913. It provides three forms of redress for people who can demonstrate a right to restitution: restoration of land lost, an award of alternative state owned land or financial compensation.

The Development Facilitation Act, 67 of 1995: The DFA was seen as an interim measure to establish and test a series of changes to the system of land development and planning. Important characteristics of the law which impact on tenure are 'fast track' procedures administered by provincial development tribunals for dealing speedily with important development projects and a section which empowers the minister to make regulations for the upgrading of tenure and services in urban informal settlements.[1]

New provincial planning laws: Several provinces have begun to draft their own planning laws, led by the two non-ANC provinces of KwaZulu-Natal and the Western Cape. Legal rationalization, the management of land use and development and the determination of roles and responsibilities at the three spheres of government are among the issues addressed in these laws. The law reform challenge arises from the numerous overlapping and contradictory laws still in place in each province. Although planning and development control laws and regulations in bantustans and urban black townships were only minimally effective, their links with land tenure and land administration make it extremely difficult to simply repeal them.

The Prevention of Illegal Eviction from and Unlawful Occupation of Land Act, 19 of 1998: This law introduces procedures for dealing with illegal occupation of land that are 'just and equitable' and aims to ensure that eviction takes place in accordance with the law. It also outlaws the practice of individuals charging landless people a fee for the opportunity to settle on land that is not that individual's land to offer. It is too early to assess the impact of this Act, although it is anticipated that the legal relationship between landowners and illegal occupiers of land will fundamentally change.

and development' projects (NDPC and DLA, 1999). In 1999 the national development and planning commission (NDPC), established in terms of the DFA, fulfilled its mandate to advise government on transforming the inherited legal framework for the development and planning of land. A national spatial planning law and white paper on spatial planning and development are currently being developed. Three additional legislative initiatives in the land occupation, land development and planning arenas have a bearing on urban tenure security (see Box 10.4).

1 As well as the initial ownership provision referred to above.

Impact on urban tenure: The legal and policy framework in practice

Responsibility for the land reform programme, including tenure reform, lies with the national department of land affairs (DLA), while the department of housing is responsible for the housing programme. The DLA has focused its concerns in rural areas while housing's implementation has tended to favour the delivery of new housing. Urban tenure reform tends to sit uneasily between the two departments and has thus suffered from policy and implementation neglect. Nevertheless, it is the housing programme that has had the biggest impact on urban tenure security, while the land reform programme is mainly rural in application, although there are large numbers of former homeland towns in which land rights are unclear, overlapping and at times conflictive. Chapter 13 provides more detailed insight into the Greater Nelspruit case, an example of this situation.

The government housing programme has made the largest impact on human settlements in the last five years (CSIR, 1999). The housing capital subsidy scheme is certainly the most significant and, arguably, the only instrument for urban tenure reform (Development Works, 1999a). In Chapter 12, Cross argues that tenure security is part of the housing package, and something of an afterthought. However, with regard to urban tenure, it has two main failings. The first is the promotion of individual ownership over other forms (points made in Chapters 11, 12 and 13) and the second is an emphasis on new, vacant land or greenfield development projects to the detriment of informal settlement upgrading. Chapters 11 and 12 indicate how informal settlement upgrading works in the context of the housing framework. Both of these factors are attributable in part to the dominance of the developer-driven, project-linked subsidy option, which is led by the formal private sector. These shortcomings are considered in greater detail below.

The promotion of individual ownership

Although housing subsidy policy may have been designed to create a variety of tenure forms, individual ownership is by far the main form of tenure delivered. By March 1999 over 1 million subsidies had been approved and over 745,000 housing units were complete or under construction (CSIR, 1999). The project-linked subsidy option is the most common and it delivers individual ownership. For example, in Gauteng, a comparison of the individual ownership project-linked subsidy with the institutional subsidy (which delivers alternative forms of tenure including institutional rental and cooperative tenure) demonstrated the predominance of the former at 77 per cent (150,000 houses) of all subsidies, with alternative tenure under the institutional subsidy at only 2 per cent or 4000 units (Narsoo, 2000).

Perceptions also play an important role in the lack of variety in tenure options being delivered. Many people perceive ownership as a right (demonstrated for example by the universal call for ownership of public rental stock by civic organizations) and a question of redress (evidenced in the policy statements in the land policy white paper which promote doing away with permits). The constitution clearly establishes that people or communities whose land tenure is legally insecure as a result of past racially discriminatory laws or practices are entitled to legally secure tenure or comparable redress (Republic of South Africa, 1996, section 25(6)). These perceptions and the constitutional provision have their roots in the historical denial of ownership to most South Africans and in the experience of the relationship between non-ownership forms of tenure and apartheid control of land rights and urban insecurity.

On the other hand, the options for alternative forms of tenure are not extensive. With the move away from apartheid forms of tenure (which included permits to occupy and deeds of grant), South African law currently makes available individual and group ownership and rental. The DFA contains an instrument for 'initial ownership' that entitles the rights holder to use and occupy land as if he or she is the owner. It can be converted to ownership at a later stage, entailing registration. However, initial ownership has not been used, possibly due to broader problems being experienced with the land development procedures in the Act. A proposed bill on land rights aimed to create statutory rights in former homeland and trust areas on state-owned land. Blanket protection and security of tenure would be granted upon enactment of the legislation, and registration of the rights would only occur in those instances where a demand for titling was expressed (anticipated instances would be situations of conflict or high potential investment areas). Investigations were under way to extend the bill's coverage to former homeland towns (Development Works, 1999b–d). However, at the time of writing, this initiative appeared to have fallen victim to the politically charged question of the role of traditional authorities in land matters. The transfer of houses programme (see Box 10.1) is currently raising the issue of the need for alternative forms of tenure, such as family title. This need has arisen from the difficulty of establishing clear entitlement, as a result of innumerable informal changes in tenancy and a variety of occupancy patterns resulting from overcrowding, including the occupation by more than one family of a single unit. When it comes to the implementation of alternative forms of tenure it is really only social housing, under the aegis of the institutional subsidy, which provides any alternatives to individual ownership. Social housing tenure forms include cooperative tenure and rental. However, these initiatives are not delivering at a large enough scale, are seldom in informal settlement contexts and do not accommodate 'the poorest of the poor'. Nevertheless, renewed openness to alternative forms of tenure is evident in recent ministerial and senior management statements to the public.

An inflexible housing framework which is undeveloped regarding regularization

In South Africa the housing framework remains relatively undifferentiated, despite a complex empirical reality, a point emphasized by Huchzermeyer in Chapter 11 on the intervention record. The standardized housing response leaves little room for local alternatives, whether defined municipally or by urban informal communities. In addition, rather than being driven by a rights-based approach (as is the case in the rural context where tenure is clearly articulated as a right of historical redress), the rationale for urban settlement intervention is rapid housing delivery at a scale to reduce apartheid's backlog and address increasing urban population growth. Tenure security is delivered as a by-product of a formal housing delivery intervention directed at this 'numbers game'.

There is little evidence of alternatives to the housing subsidy scheme or its relatively uniform application. Although the housing policy endorses a wide variety of delivery approaches, implementation tends to be biased towards the development of new housing and vacant land. This delivery emphasis supports the logic of speedy delivery at large scale in pursuit of the '1 million houses in five years' target set by the ANC in its 1994 election manifesto. This trend tends to leave the question of securing existing conditions of insecure tenure out in the cold, with the exception of those subsidy projects that constitute informal settlement upgrading. (Huchzermeyer and Cross address the question of informal settlement upgrading in greater detail in Chapters 11 and 12, respectively.) Informal settlement intervention, via the subsidy scheme, applies the vacant land development logic. It replaces shack settlements with formally laid out and standardized townships, prohibiting and discouraging the construction of formal dwellings prior to the standardization of the layout (Huchzermeyer, 2001b). An unambiguous and coherent policy on regularization is absent (Royston, 1998a). The housing policy has resulted in an emphasis on product – a package of land, tenure, infrastructure and 'top structure' (a starter house) – with some exceptions, such as the serviced stands delivered as a component of the Gauteng Province's incremental approach to housing.

If informal settlement upgrading was to be more firmly located on the policy and implementation agendas, what would be the implications of in situ upgrading for urban access and integration? The settlement patterns that have emerged as a result of South Africa's urbanization process are diverse but characterized in many instances by informality and peripheral settlement location – an historical legacy with which tenure interventions are confronted. The confirmation of de facto rights assumes an in situ approach, which runs the risk of cementing inherited, inaccessible, inequitable and inefficient patterns of settlement. The dilemma pits a long-term planning and economic rationale against social justice and redress. Is in situ tenure regularization viable in peripheral places from the perspectives of sustainable development, public and household investment and economic activity? On the other hand, what constitutes a better location in situations where social capital is invested where people currently are? In terms of the spatial integration principle, public investment in services is likely to occur in 'good' locations. If tenure upgrading is coupled with the delivery of services, then people who cannot afford services will be squeezed out. In Chapter 12 Cross demonstrates how formal development can prove to be unsustainable in terms of household income.

If securing urban tenure was to be more centrally located in the housing policy, then the implications for the registration system would need to be addressed. The exactitude of, and high regard for, South Africa's deeds registration system is widely acknowledged. However, in the absence of a rights-based, as opposed to titling, approach, its highly centralized nature would test its responsiveness to a major programme of regularization. The impact of the proposed Registration Facilitation Bill on contexts of urban settlement tenure reform, which would create an electronic registration system accessible by local authorities, could be important.

MAJOR CHALLENGES IN URBAN TENURE REFORM

A major set of urban tenure challenges lies in reforming or amending the policy and legal framework and in adapting its application to implementation. The following points indicate directions for change:

- a more holistic approach to urban settlement development, especially policy and procedures for informal settlement upgrading, including securing tenure, and for addressing informal rental, two significant gaps in the existing framework;
- greater attention to alternative forms of tenure, including statutory forms of tenure;
- a more varied and important policy response and menu of delivery options to cater for diverse needs and affordability levels, than the project-linked, private ownership subsidy route (the people's housing process has not, as yet, been evaluated at sufficient scale);
- rationalizing the legal maze inherited from the apartheid regime and, in some respects, extended by new planning law; and
- simplifying and decentralizing the registration system.

In addition, greater clarity is required on institutional relationships if more progress is to be made in the realm of tenure security. The first component of this challenge lies in getting the relationship right between the national housing and land affairs departments on the urban tenure question. The second institutional challenge relates to the capacity of local government to undertake responsibility for housing delivery. Local authorities are faced with managing the vexed question of land-related conflict, especially 'competing jurisdictions' between informal allocators (including traditional authorities) and formal, programmed land release and housing development strategies of government. In Chapter 14, Davies and Fourie propose a land management system for local authorities, which could assist in the management and upgrading of informal settlements. Most municipalities will also be challenged by having to incorporate urban tenure concerns into strategic planning processes (called integrated development planning) and to take on the body of related responsibilities and opportunities that flow from upgrading, including developing, extending and maintaining the delivery of services, a rating system and a land use management system.

References

Arrigone, J (1994) *Self-help Housing and Socio-economic Development*, Development Bank of Southern Africa, Policy Working Paper 13, Midrand

Baumann, T (1998) *South Africa's Housing Policy: Construction of Development or Development of Construction*? Report to Homeless International, Cape Town

Berrisford, S (1999) 'Urban Legislation and the Production of Urban Space in South Africa', Paper presented at the international conference *South to South: Urban–environmental Policies and Politics in Brazil and South Africa,* Institute of Commonwealth Studies, London, 25–26 March

Crankshaw, O (1993) 'Squatting, Apartheid and Urbanization of the Southern Witwatersrand', *African Affairs*, Vol 92, No 366, pp31–61

CSIR (Council for Scientific and Industrial Research) (1999) 'The State of Human Settlements: South Africa 1994–1998', Draft report prepared for the Department of Housing, Pretoria

Development Works (1999a) *Tenure Security in South Africa: A Critical Review of the National Legislative and Policy Framework and the Implications for Intervention,* Report submitted to the French Institute of South Africa, Johannesburg

Development Works (1999b) 'Recommendations on R293 Towns and Conditions of Unregistered and Informal Tenure', *Urban Tenure Policy Research: Report 1,* Submitted to the Department of Land Affairs, Pretoria

Development Works (1999c) 'R293 Provincial Scan', *Urban Tenure Policy Research: Report 2,* Submitted to the Department of Land Affairs, Pretoria

Development Works (1999d) 'Greater Nelspruit Case Study', *Urban Tenure Policy Research: Report 3,* Submitted to the Department of Land Affairs, Pretoria

Emdon, E (1993) 'Privatisation of State Housing – with Special Focus on the Greater Soweto Area', *Urban Forum*, Vol 4, No 2

Engelbrecht, C, Chawane, H, Gallocher, R, van Rensburg, G, Mkhalali, X, van der Heever, P and Browne, C (1999) *Upgrading Informal Settlements: a Brazilian Experience,* Gauteng Department of Housing, Johannesburg, 1 July

Hindson, D and McCarthy, J (1995) *Urban Land Management, Regulation Policies and Local Development in Kwazulu-Natal,* Groupement de Recherche Interurba, CNRS, Bordeaux

Huchzermeyer, M (1999) 'The Exploration of Appropriate Informal Settlement Intervention in South Africa: Contributions from a Comparison with Brazil', unpublished PhD thesis, Department of Sociology, University of Cape Town, Cape Town

Huchzermeyer, M (2001a) 'Consent and Contradiction: Scholarly Responses to the Capital Subsidy Model for Informal Intervention in South Africa', *Urban Forum*, Vol 12, No 1, pp71–106

Huchzermeyer, M (2001b) 'Housing for the Poor? Negotiated Housing Policy in South Africa', *Habitat International*, Vol 25, pp303–331

Khan, F (1995) 'Rental Housing in South Africa', CCLS (UDW) Housing and Urban Development Project, Working Paper 1, Centre for Community and Labour Studies, University of Durban-Westville, Durban

Kok, P and Gelderblom, D (1994) *Urbanization: South Africa's Challenge,* 2, HSRC Publishers, Pretoria

Mashabela, H (1990) *Mekhuku: Urban African Cities of the Future,* South African Institute of Race Relations, Johannesburg

Narsoo, M (2000) 'Critical Policy Issues in the Emerging Housing Debate', Paper presented at *Urban Futures Conference,* Johannesburg, July

NDPC (National Development and Planning Commission) and DLA (Department of Land Affairs) (1999) *Green Paper on Development and Planning,* Pretoria

Republic of South Africa (1994) *White Paper: A New Housing Policy and Strategy for South Africa,* Government Gazette 354,16178, Pretoria

Republic of South Africa (1995) *Development Facilitation Act,* Act 67 of 1995

Republic of South Africa (1996) *Constitution of the Republic of South Africa,* Act 108 of 1996

Republic of South Africa (1997) *White Paper on South African Land Policy,* Department of Land Affairs, Pretoria

Robinson, P, Sullivan, T and Lund, S (1994) *Assessment of IDT's Capital Subsidy Scheme,* Consultants' report to the Independent Development Trust, Johannesburg

Roux, T (1999) 'Something Less than Ownership: Extending Security of Tenure in Rural South Africa', Paper presented at the International Research Group on Law and Urban Space/Centre for Applied Legal Studies workshop on *Facing the Paradox: Redefining Property in the Age of Liberalization and Privatisation,* Johannesburg, 29–30 July

Royston, L (1998a) *Urban Land Issues in Contemporary South Africa: Land Tenure Regulation and Infrastructure and Services Provision,* Development Planning Unit Working Paper, No 87, London

Royston, L (1998b) 'South Africa: the Struggle for Access to the City in the Witwatersrand Region', in A Azuela, E Duham and E Ortiz (eds) *Evictions and the Right to Housing,* International Development Research Centre, Ottawa

Royston, L (1999) 'The Socio-political Dynamics of Urbanization in South Africa and its Prospects', Paper presented at international conference *South to South: Urban–Environmental Policies and Politics in Brazil and South Africa,* Institute of Commonwealth Studies, University of London, London, 25–26 March

Stevens, R (1999) 'Technical Visit to Lima (Peru), South America', Gauteng Provincial Government Department of Land and Housing, Report, Johannesburg, 22 January

Urban Foundation (1991) *Policies for a New Urban Future: Informal Housing,* The Urban Foundation, Johannesburg

Xaba, S and Beukman, R (1999) 'The Upgrading of Informal Tenure in South Africa: the Policy Gaps and Legislative Challenges', Paper presented at the international workshop on *Tenure Security Policies in South African, Brazilian, Indian and Sub-saharan African Cities: A Comparative Analysis,* University of the Witwatersrand, Johannesburg, 27–28 July

Chapter 11

Evaluating Tenure Intervention in Informal Settlements in South Africa

Marie Huchzermeyer

Informal settlement intervention in South Africa is structured through the current national housing policy, which entitles low-income households to a one-off capital subsidy. Developed by the Urban Foundation in the late 1980s and mainstreamed through the Independent Development Trust (IDT) capital subsidy scheme in the early 1990s, this approach is geared towards the delivery of standardized serviced sites. For the implementation of the development concept promoted by the Urban Foundation, the IDT was set up in 1990 with a R2 billion government grant dedicated to social upliftment. Through the IDT, 100,000 households (primarily resident in informal settlements) were to benefit from capital subsidies in the form of serviced sites.

Currently, the individual capital subsidy covers the cost of a serviced plot of land and an incremental or starter house. The housing delivery package is predominantly associated with individual freehold tenure. Through this intervention, informal settlements are replaced with standardized and highly individualized residential developments. For informal settlement residents, such intervention is the culmination of a drawn-out endeavour to secure tenure. In most cases it implies the granting of temporary rights to occupation, usually through the numbering of shacks and the imposition of certain controls that enable eventual relocation to a formally laid out development site. Only in exceptional cases has formal development taken place on or near the invaded land. Does this policy represent an adequate government response to the tenure requirements of the urban poor?

This chapter draws together findings from the South African literature of the 1990s that has sought to address this question. It reviews the functioning of informal settlements and, more specifically, the management of land tenure, prior to intervention, as well as the outcomes of the standardized intervention. This literature is by no means comprehensive and gaps and biases that have bearing on the land tenure problematic are highlighted.

THE FUNCTIONING OF INFORMAL SETTLEMENTS PRIOR TO CAPITAL SUBSIDY INTERVENTION

Informal settlements are characterized by insecure tenure. The extent of this insecurity, or the intensity of the threat of eviction, varies over time. Though driven by the scarcity and cost of formal housing, the invasion of urban land responds to political conditions. In South Africa, the period surrounding the 1994 elections was favourable to land invasion. Typical of this period was the formation of the Wiggins settlement in Cato Manor, Durban. It was based on the ambiguity of the moral and political authority on the eve of democracy, the unpreparedness of any party to respond decisively at the risk of losing votes, and awareness by the invaders of the waning authority of the state (Gigaba and Maharaj, 1996). However, this opportunity for a successful organized invasion in South Africa was short-lived. The demolition of 400 shacks at Cato Manor is among the evidence suggesting that the new government, 'fully aware of the legitimacy of its political and moral authority, was prepared to be decisive about invasions' (ibid, p233). South African research literature has yet to document the range of post-election (1994) land invasion processes and their contexts.

Where the invaded land is embedded in existing townships, the localized level of politics has been harnessed for the land invasion process (Crankshaw, 1996, p55). This approach, typical of the 1980s and 1990s, requires collective organization and lobbying, which in itself gives individuals a sense of security and control over the process. A contrasting process to invasion is that of gradual occupation by individuals, as described in Ardington's (1992) study of the Canaan settlement in Durban around 1990. There, desperate individuals took advantage of vacant land at a suitable location. Seeming tolerance by authorities enticed further individuals to join. Previous experience of political violence and the resulting displacement from their homes discouraged these invaders to organize collectively. Only the issuing of eviction notices, the first concrete threat to security of tenure, led to the formation of a settlement committee. However, community cohesion around the common threat of eviction (in this case removal to an inconvenient relocation site) disintegrated once it became apparent that their resistance to eviction had failed. Once again, they had to make individual decisions about their residential future.

The process of securing tenure from individual to collective action has been described in more detail for other settlements (Adler, 1994; Makhatini, 1994; Royston, 1998). In an analysis of the invasion process at Cato Manor (Makhatini, 1994), three distinct stages have been identified. The first stage was camouflaged or hidden squatting, comprising the initial invasion by individuals and the formation of social cohesion in response to a shared threat. Political party mobilization was found to be absent at this stage; mainly socio-economic arguments were used by authorities in the organized lobbying. The process of establishing 'codes of behaviour and degrees of sacrifice' renders this stage prone to conflict (ibid, p11). At the same time, economic opportunities are created through clandestine building activity (much of which occurs at night) and through an increasing demand for basic household commodities. This stage ends with a first step in government intervention: the official numbering of shacks and an associated mandate to the settlement organization to prevent the construction of additional shacks.

The numbering of shacks signals a degree of tenure security to the residents and this leads to the second stage, namely open squatting. This is characterized by a social distinction between legitimate insiders and the newcomers who present a threat to the official deal. Due to the moratorium on building construction, newcomers are diverted to new 'camouflaged' land invasions, to which local informal employment and construction-related economic opportunities are also diverted. Settlement leadership at the stage of open squatting needs to be sophisticated in order to negotiate for service delivery. The third stage is one of organization and consolidation, characterized by leadership struggles around securing permanent tenure and development deals. It may also be characterized by divisions arising from different development preferences, from party political mobilization and 'the emergence of overtly aligned power structures' (Makhatini, 1994, p16). At the time of Makhatini's study, the Cato Manor development was in this third stage.

It may take several decades to progress from cautious invasion by individuals to organized consolidation via an official process. An example of this is the Piesang River invasion in Durban in the 1960s. Community organization followed only in the 1980s, and there was a drawn-out process of formalization spanning most of the 1990s (Huchzermeyer, 1999). It is relevant to examine how demands on land are managed prior to the granting of permanent tenure rights and the imposition of formal layouts.

Urban informal settlements in South Africa vary considerably in size and density. Very little research has addressed the ways in which space is managed under these different scenarios. In an unpublished study (Dewar and Wolmarans, 1994) five informal settlements in the Western Cape were compared: Bloekombos and Norodhoek Site Five in peri-urban areas of Cape Town; Sun City and Waterworks in small rural towns of the Western Cape; and the historical case of Crossroads in the late 1970s, adjacent to the Cape Town township of Nyanga. At the time of the research all the settlements were in the stages between 'open squatting' and 'organization and consolidation'. In line with a general bias in informal settlement research in South Africa, none of these case study areas was in close proximity to formal townships, other than the now largely eradicated Crossroads settlement, which the researchers accessed through historical documentation of the situation around 1979. Nevertheless, the study addresses an interesting range of informal spatial and organizational responses to the characteristics of the invaded land. The underlying drive of the research is a positive view of people's initiative, thus giving insight into collective coping mechanisms spanning most aspects of the living environment.

Shack densities in the five case study settlements range from 38 to 220 dwelling units per hectare. Each settlement is confronted by a continuous demand to accommodate additional households. Despite official sanctioning of additional shack construction, the flexible management of space responds to the urgency for accommodation. Where land is scarce, thereby not allowing for expansion on the settlement edges, internal subdivision of space has led to extreme densities of dwelling units. This places a particular challenge on the collective management of the following: access; the disposal of refuse, grey water and excrement; the accommodation of private and collective outdoor activities; the demarcation of plots; the accommodation of communal facilities (ranging from sports fields, communal grazing and wood lots to churches and community halls); and the location of commercial activity. The study documents innovative, though precarious, solutions within the settlement layouts, which are responses to the specific conditions of the site and surroundings.

The question of tenure in informal settlements is directly linked to social organization and associated control over resources. In the Western Cape study reviewed above, the plot allocation process is associated with the democratic structures of community or civic organization. Control by the civic structure over plot allocation is described as a noteworthy approach. It was required that newcomers be introduced to the committee by a resident. They were then screened and offered a choice of sites, where possible in the vicinity of their sponsors and after negotiation with the neighbouring residents. Precise boundaries were negotiated between neighbours, with the committee intervening in the case of a dispute. Where temporary rights to land had been officially granted, security of tenure was threatened should newcomers be admitted. However, the study documents ongoing densification despite official sanctioning. It appears that community organizations were compelled to accommodate newcomers because established households were gradually ceding portions of their private space to newly arriving friends or relatives. The study further documents the renting of rooms and individual beds – or lodging.

The reality of ongoing settlement growth and lodging, which creates a degree of population fluidity, raises difficult questions around the official intervention approach. It questions the practice of shack numbering and freezing of settlements as an early step in a development process. This process requires settlement stabilization in order to eventually tie the population into individual freehold tenure and service charge payments. Research in the Durban functional region illustrates how the informal tenure practices 'have adapted to mobility and, to a considerable extent, represent portable land rights' (Cross, 1994, p187). They comprise 'systems of relative social rights, rather than systems of property rights' (ibid). With reference to the obstructions of formal private tenure for low-income users, the study recommends the formalization of informal tenure, thus 'a formal version of what impoverished shack communities produce for themselves' (ibid, p188). It is argued that, while unpopular with planners, this approach 'offers flexible site access to its users and can usually secure their rights in terms of accepted understandings against anything short of violence or government intervention' (ibid, p187). In order to understand this concept it is necessary to briefly review the characteristics of informal tenure in the Durban area.

In the 1970s and 1980s rent tenancy was common in informal settlements in Durban. This comprised the renting of rooms and backyard structures, as well as site rental. Landlords were making profits or securing business clientele, and shack lords were securing political or personal support. In the late 1980s and early 1990s, these exploitative practices gave way to a 'powerful social movement against the practice of paying rent for access to land' (Cross, 1995, p31). The structures of the 'mass democratic movement', including civic committees, came fully into the open and took up control in urban informal tenure. This comprised a significant shift away from the ethic of private land tenure (albeit rental and not freehold), towards a land ethic 'closer to communal land holding in rural areas' (Cross, 1994, p180). Thus the entry process of sponsorship and screening, as described for the Western Cape settlements above, has parallels in rural tenure systems. It has been pointed out, however, that tenancy relationships have endured in some settlements, where entry through the purchase of sites from the leadership continues to be practised. Clandestine selling of sites in settlements controlled by democratic civic organizations has likewise been documented (Cross, 1995, p34).

Within the widespread system of sponsorship and screening for access to land in informal settlements, concern has been raised over the sidelining of single people and, in particular, single women and women with young children (Cross, 1995). Such women resort to a range of options. First, they may enter into 'clientship arrangements with established householders' (ibid, p35), though this often results in suspicion from established female residents. Second, they may attempt to enter into new settlements that are still in the recruitment stage. As a third option, they may lodge within an informal settlement, thereby being in a position to collect the necessary local information that may eventually lead to the allocation of a site (ibid). Thus the finding in international literature on informal settlements that lodging ranks lowest in the residential hierarchy and is largely occupied by women (Volbeda, 1989; Yapi-Diahou, 1995), applies to some extent to the South African situation.

Despite inequalities inherent in the informal tenure system, it is evident that the civic movement had a positive impact on conditions in informal settlements by reducing the cost of living through the eradication of exploitative rental arrangements. Cross implicitly recognizes the contribution of the civic movement in 'facilitating access to housing' (1994, p187) through the eradication of rental tenancy. She goes further to warn that the imposition of private freehold titles through the development process may 'contribute to a loss of equity and a re-emergence of rent tenancy' (ibid). This has parallels with Mayekiso's (1996) concern about new forms of exploitation made possible through the privatization of public housing in townships. This would lead to the reintroduction of exploitative landlord–tenant relationships in the backyard shack situation.

In turning to the evaluation of the formal tenure intervention process, it is necessary to note that the impact of commercialization has not been adequately researched. The formal tenure intervention process has been induced by close on a decade of issuing freehold title through the capital subsidy system in South Africa. Nonetheless, as will be shown, the discrepancy between individual freehold title and the reality on the ground is frequently alluded to in the literature.

Subsequent to capital subsidy intervention: Evaluations of the intervention record

The literature evaluating informal settlement intervention is relatively cohesive. It deals with only one approach, namely that of replacing informal settlements with standardized residential developments. Where such development occurs on the invaded land, the term 'in situ upgrading' is applied. The informal settlement intervention record in South Africa has been evaluated predominantly from the perspective of the private sector. Former Urban Foundation researchers and practitioners have undertaken influential evaluations, in defence of the standardized intervention approach through the capital subsidy and the imposition of freehold tenure. Few evaluations have taken a more independent perspective, portraying the reality of the beneficiaries and the experience of community organizations.

In situ upgrading in South Africa is carried out through the housing subsidy system, and is therefore a product-oriented approach. Though officially termed 'incremental', the South African housing policy sets out to deliver a standardized product, to be improved incrementally by the beneficiary household. Due to this product orientation, the difference between in situ upgrading and greenfield development in the housing policy discourse is merely one of location, and not of approach. Both in situ upgrading layouts and greenfield development layouts are standardized through conventional infrastructure norms and regular plot sizes. An evaluation of informal settlement intervention in South Africa found that 'Once an informal settlement has been upgraded in situ, it does not differ fundamentally from a settlement where housing has been delivered on an incremental basis' (McCarthy et al, 1995, pp2–3). Exceptions have been the in situ interventions in hilly areas in and around Durban, where the topography does not lend itself to a regular layout. Attempts have been made to base official layouts on the existing pattern of land occupation. However, the intervention remains product oriented with a focus on standardized services, houses and individual freehold titles. The case of the Kanana settlement in Sebokeng illustrates the struggle of an organized informal settlement community to have its unofficial layout recognized and formalized without the disruption of the informally established settlement pattern (People's Dialogue, 1997; Huchzermeyer, 1999).

This has been the approach in South Africa throughout the 1990s. It contrasts with what is usually referred to internationally as 'in situ upgrading', namely the introduction of infrastructure, community facilities and secure tenure with minimal disruption to the physical and social fabric of the settlement. This discrepancy between international and local practice has been highlighted through the case of Bester's Camp in KwaZulu-Natal. In the late 1980s the Urban Foundation oriented its pilot intervention in this settlement to internationally acclaimed practice: building a sensitive 'bottom–up' approach in direct response to community-based priorities. However, in an internal contradiction, the Urban Foundation came to promote the standardized capital subsidy as the sole mechanism for informal settlement intervention.

With the advent of the IDT capital subsidy scheme, the standardized intervention approach, including individual freehold title, was imposed on the upgrading of Bester's Camp. This overrode the priorities that had been developed from within the organizational structures of the community. In an evaluation of this intervention it is noted that by 'linking the payment of subsidies to the delivery of legal tenure, the project became strongly product driven by the imperative to deliver serviced sites and legal tenure' (van Horen, 1996, p24). Thus in situ upgrading in Bester's Camp came to mean the delivery of a standardized product.

Thus the term 'upgrading' in South Africa is applied to extensive residential developments. An example of this is the Integrated Serviced Land Project (ISLP) in Cape Town (see Tomlinson, 1996, p59), which has relocated large numbers of informal settlement households to its peripherally situated relocation sites. Where this project has 'upgraded' land invasions (for example, Millers Camp and KTC) a roll-over procedure has been applied; the invaded land is cleared, a selected portion of the population is relocated to peripherally located formal sites and the vacated land is developed according to standard greenfield procedures. The remaining portion of the original population is then formally resettled on the developed sites.

Due to this standardized approach, evaluations of informal settlement intervention in South Africa have not differentiated between in situ upgrading and greenfield development. Thus McCarthy et al (1995), although setting out to evaluate 'informal settlement upgrading and consolidation projects', include in their case studies both greenfield and in situ upgrading projects. The conclusion of the study does not differentiate between what was found in the in situ upgrades as opposed to the greenfield situations. The same limitation applies to the extensive USAID (US Agency for International Development) funded evaluation of the capital subsidy intervention by the Centre for Policy Studies (Tomlinson, 1995a, b, 1996, 1997a, b, 1998).

A number of post-development evaluations promote the product-oriented capital subsidy approach, despite giving evidence of its failure (particularly on aspects of the tenure intervention). These evaluations tend to be biased towards the perspective of private sector consultants and developers and do not grasp the reality of the informal settlement residents. They represent problematic views and assumptions on the functioning of property markets in low-income settlements. These are also the assumptions on which the standardized capital subsidy approach is based, with its emphasis on individual freehold tenure.

A key assumption in these evaluations is that imposing individual freehold titles can create a property market equal to that of a middle class suburb. Thus Marais and Krige, in the conclusion to their post-development survey of the Freedom Square settlement in Bloemfontein, praise the transformation 'from an informal settlement ... to a suburb' (1997, p189). This is despite evidence in the same study of significant shortcomings in the development approach, particularly with regard to the functioning of the property market. The study found that 'a number of stands have been vacated [without selling] after the sites were registered in the names of the owners' (ibid, p185). Botes made the same finding, which was identified as a 'financial threat to the development because the capital subsidy guaranteed by the IDT would only be paid once ownership of the land had been transferred' (Botes et al, 1996, p457). Their explanation for the abandoning of sites after development is that the capital subsidy, in the absence of any monetary contribution by the beneficiaries, did not sufficiently entrench a sense of ownership. They therefore recommend the imposition of a nominal fee.

Another explanation for the abandoning of sites presented by the private sector evaluators is that the residents do not understand 'the value of their sites' (Marais and Krige, 1997, p185). In an evaluation of several capital subsidy developments, McCarthy et al (1995, p22) add that this lack of understanding 'is especially true of the less educated'. The assumption behind this view is that the exchange value of an upgraded site is equivalent to the subsidy investment. This ignores the fact that the exchange value of property is primarily determined by demand and the economic power of that demand. More in-depth examinations of the post-development property market have found selling prices ranging from one-tenth to a half of the capital subsidy amount, not significantly higher than selling prices in informal settlements prior to the capital subsidy intervention (Barry, 1995; van Horen, 1996).

Rather than recognizing the limited economic power of those likely to buy into capital subsidy developments, the private sector-oriented evaluators associate the absence of a 'viable market in housing within the [upgraded] informal settlements' with 'the lack of availability of a housing market in other areas to which a family might wish to move' (McCarthy et al, 1995, p57). What these evaluators are in actual fact promoting is the displacement of original capital subsidy beneficiaries by higher income groups able to purchase the developed sites for at least the capital subsidy amount. Although this has not been adequately researched, one may assume that it is the peripheral location, the vastness, monotony and possibly the stigmas attached to these developments that have prevented interest by higher income groups.

It is in the McCarthy et al study (1995) that the private sector orientation is particularly explicit. The study repeatedly promotes an increased stake for the private sector (ie profit-making) in the upgrading process. Thus it lists the private sector alongside community leaders, professionals and the local or provincial authorities as the essential actors to make upgrading a success (ibid, p41). The study then argues for a modified role for the local authority, with 'much stronger involvement of the private sector' (ibid, p53–54). The interests of the private sector are further promoted through the statement that '[u]pgrading can unleash the huge consumer markets in informal settlements. The introduction of electricity, for example, encourages the consumption of "white goods" [meaning stoves, refrigerators, freezers and washing machines], kitchen appliances, television sets, etc' (ibid, 1995, p77).

A correlation is made between the imposition of individual freehold titles and the nurturing of capitalist values. The study maintains that the capital subsidy intervention 'has led to the accumulation of fixed assets amongst the poor that would not otherwise have materialized' (ibid, p82). A comprehensive analysis of the diverse processes of poverty and survival, and the reliance by the poor on a diverse range of social as well as physical assets, would contradict the position represented in the private sector-oriented studies. However, in-depth studies on the ongoing processes of poverty subsequent to capital subsidy intervention are lacking in the evaluation literature.

In the earlier part of this chapter, it was pointed out that the management of a responsive and flexible informal tenure approach, prior to capital subsidy intervention, is based on the level of collective community organization or representation. Decisions over the adjustment of layouts or entitlements to land were found to rest with civic organizations. It is worrying that most post-development evaluations do not consider the views of community or civic organizations. Views from the growing Homeless People's Federation movement, particularly prevalent in urban informal settlements, are likewise ignored.

The extensive and much quoted evaluation by the Centre for Policy Studies omits the views of community organizations, though factually covering those of implementers (Tomlinson, 1995a), developers (Tomlinson, 1995b), individual beneficiaries (Tomlinson, 1996), financial institutions (Tomlinson, 1997a), national and provincial legislators (Tomlinson, 1997b) and local governments (Tomlinson, 1998). This omission points to a widespread belief among development practitioners that civic or organized community activity is irrelevant or even problematic. Thus Tomlinson refers to government as 'perhaps the only institution capable of representing beneficiary interests' (1996, p52). Her study further argues that:

> *it is neither plausible nor realistic to assume that developers, civics or others with an interest in particular forms of housing delivery will cease pursuing their interests in a way which disadvantages beneficiaries, if there is no incentive for them to do so, or sanction if they do not* (ibid, p51).

Community representatives are therefore portrayed as exploitative and self-seeking. The relevance of civic or community organizations in the development process is evident from a factual evaluation of the Mandela Village upgrade north of Pretoria (Hall et al, 1996). This study acknowledges that the organizational structures of the African National Congress (ANC) and the South African National Civic Organization (SANCO) led to the abolition of payment for occupancy rights in the informal settlement and formed the basis for strong community cohesion. On this and other findings, the evaluators base their support for the emerging housing support initiative of the national department of housing, which places responsibility for the management of house construction with the residents. This position therefore challenges the dominant developer-driven housing delivery model. The study, however, does not make recommendations with regard to tenure.

The difficult position that community organizations occupy in relation to the current tenure intervention model is evident in the case of Canaan in Durban (Ardington, 1992), reviewed earlier in this chapter. The study indicates the social impact from the controls that are imposed with the granting of temporary tenure rights in preparation for the capital subsidy intervention. It points to the strain this places on the integrity or legitimacy of community representatives. Although elected by the residents, these community representatives are expected to implement the controls imposed by the authorities, rather than respond to the demands from within the community.

In the case of Canaan, the occupied land was officially classified as not upgradable for low-income residential purposes and the authorities sought a relocation site. The officials expected the settlement organization to ensure no further shack construction in the settlement. As neither the relocation site nor the moratorium on construction (prohibiting even the construction of pit latrines) met the diverse realities and needs of the residents, their social cohesion was undermined. Ardington argues that a

> *community which had come together for a multiplicity of reasons could not reasonably be expected to adopt a unanimous view on an area to relocate to and the type of housing which should be acceptable* (1992, p23).

The study concludes that due to the insensitive model of intervention,

> *Canaan, which appeared well on the way to developing a strong civic authority, is weak, divided and unable either to press its own demands, or react to those of the authorities* (ibid).

In the case of Bester's Camp, the invaded land was officially deemed appropriate for in situ upgrading. However, as argued in van Horen's evaluation, the IDT imposed approach through the capital subsidy, inadequately considered 'the pre-existing de facto system within the settlement' (1996, p36). While the project staff had attempted to root the intervention, both physically and socially, in the de facto situation, 'the time and financial pressure associated with tenure delivery ... did not permit successive iterations which would have more accurately reflected ongoing changes in the settlement' (ibid, p23). The study thus emphasizes the discrepancy between the de facto situation and the imposed title registration system (and its cost) as the de jure solution. Also reviewing the Bester's Camp upgrade, Merrifield et al (1993, p10) mention the exploration of tenure alternatives by the project staff, who 'questioned the sustainability of a first world land registration system' and were hoping to apply a tenure regularization process that would integrate mechanisms of the informal tenure system. To date, the product-oriented framework of the capital subsidy has not allowed such tenure alternatives to be considered for the permanent legalization of informal settlements.

Conclusion

Although the evaluation literature reviewed in this chapter is by no means comprehensive, the conclusion drawn is that the standardized approach of imposing individual freehold titles through a capital subsidy intervention in informal settlements is inadequate. The standardized approach is underpinned by incorrect assumptions about the functioning of property markets, disregarding the need for flexibility in the management of land entitlements.

This chapter has contrasted the functioning of informal settlements and, in particular, the informal management of land prior to the imposition of formal layouts and individual freehold titles with the problems experienced subsequent to intervention. This contrast has indicated the need for a formal tenure mechanism that will accommodate elements of the approach that was informally practised prior to intervention through civic organizations in informal settlements. Though subtle forms of exclusion were found to be inherent in this informal practice, it does appear to form the basis of an appropriate system of entitlement. It is by exposing and debating its shortcomings in relation to fundamental human rights that the flexible tenure approach may be improved and eventually legally accepted.

However, the intervention framework, as it is currently structured through the household-based capital subsidy, does not lend itself to a flexible tenure approach that may be managed by representative community-based organizations. The exploration of alternative tenure approaches must therefore be accompanied by a search for an alternative framework to that of the product-oriented capital subsidy. This is necessary if the national housing finance is to be channelled into the adequate improvement of informal settlements.

Acknowledgements

This chapter is based on research conducted by the author in the Department of Sociology, University of Cape Town, and funded by the Division of Building Technology (Boutek) of the Council for Scientific and Industrial Research (CSIR), Pretoria.

References

Adler, J (1994) 'Life in an Informal Settlement', *Urban Forum*, Vol 5, No 2, pp99–110

Ardington, E (1992) *Buckpassing in Canaan: An Example of Authorities' Failure to Address the Needs of Informal Urban Dwellers,* Rural Urban Studies Working Paper No 24, Centre for Social and Development Studies, University of Natal, Durban

Barry, M (1995) 'Conceptual Design of a Communal Land Registration System for South Africa', *South African Journal of Surveying and Mapping,* Vol 23, No 3, pp153–162

Botes, L, Stewart, T and Wessels, J (1996) 'Conflict in Development: Lessons from the Housing Initiatives in Freedom Square and Namibia Square', *Development Southern Africa,* Vol 13, No 3, pp453–467

Crankshaw, O (1996) 'Social Differentiation, Conflict and Development in a South African Township', *Urban Forum,* Vol 7, No 1, pp53–67

Cross, C (1994) 'Shack Tenure in Durban', in D Hindson and J McCarthy (eds) *Here to Stay: Informal Settlements in KwaZulu-Natal,* CSDS, Indicator Press, University of Natal, Durban

Cross, C (1995) 'Shack Tenure in the Durban area', in C Hemson (compiler) *Land Use and Land Administration in the New South Africa,* Proceedings of a Winter School held at the University of Natal, Durban, July

Dewar, D and Wolmarans, P (1994) 'Responsive Environments: A Spatial Analysis of Five Squatter Settlements in the Western Cape', Unpublished manuscript, School of Architecture and Planning, University of Cape Town

Gigaba, M and Maharaj, B (1996) 'Land Invasions during Political Transition: The Wiggins Sagga in Cato Manor', *Development Southern Africa,* Vol 13, No 2, pp217–235

Hall, P, Napier, M and Hagg, G (1996) 'The Cart Before the Horse: Housing and Development in Mandela Village', *Indicator SA*, Vol 13, No 4, pp67–71

Huchzermeyer, M (1999) 'The Exploration of Appropriate Informal Settlement Intervention in South Africa: Contributions from a Comparison with Brazil', unpublished PhD thesis, Department of Sociology, University of Cape Town, Cape Town

McCarthy, J, Hindson, D and Oelofse, M (1995) *Evaluation of Informal Settlement Upgrading and Consolidation Projects,* Report to the National Business Initiative, Johannesburg

Makhatini, M (1994) 'Squatting as a Process: The Case of Cato Manor', Paper presented at the University of the Witwatersrand History Workshop, Johannesburg, 13–15 July

Marais, L and Krige, S (1997) 'The Upgrading of Freedom Square Informal Settlement in Bloemfontein. Lessons for Future Low-income Housing', *Urban Forum,* Vol 8, No 2, pp177–194

Mayekiso, M (1996) 'Township Politics: Civic Struggles for a New South Africa', Monthly Review Press, New York

Merrifield, A, van Horen, B and Taylor, R (1993) *The Politics of Public Participation in Informal Settlement Upgrading: A Case Study of Bester's Camp,* Proceedings of the 21st International Housing Association, World Congress, Cape Town, May

People's Dialogue (1997) *The Liberating Power of Self Reliance: People Centred Development in Kanana Settlement, Vaal Region, Gauteng Province,* Report compiled by People's Dialogue on behalf of the South African Homeless People's Federation, Cape Town

Royston, L (1998) 'South Africa: Struggle for Access to the City in the Witwatersrand Region', in A Azuela, E Duhau and E Oritz (eds) *Evictions and the Right to Housing,* International Development Research Centre, Ottawa

Tomlinson, M (1995a) *From Policy to Practice: Implementers' Views on the New Housing Subsidy Scheme,* Research Report No 44, Centre for Policy Studies, Social Policy Series, Johannesburg

Tomlinson, M (1995b) *Problems on the Ground? Developers' Perspectives in the Government's Housing Subsidy Scheme,* Research Report No 46, Centre for Policy Studies, Social Policy Series, Johannesburg

Tomlinson, M (1996) *From Rejection to Resignation: Beneficiaries' Views on the Government's Housing Subsidy Scheme,* Research Report No 49, Centre for Policy Studies, Social Policy Series, Johannesburg

Tomlinson, M (1997a) *Mortgage Bondage? Financial Institutions and Low Cost Housing Delivery,* Research Report No 56, Centre for Policy Studies, Social Policy Series, September

Tomlinson, M (1997b) *Watchdog or Lapdog? National and Provincial Legislators' Views of the New Housing Subsidy Scheme,* Research Report No 59, Centre for Policy Studies, Social Policy Series, Johannesburg

Tomlinson, M (1998) *Looking at the Local: Local Governments and Low Cost Housing Delivery,* Research Report No 63, Centre for Policy Studies, Social Policy Series, Johannesburg

Van Horen, B (1996) 'Informal Settlement in situ Upgrading in South Africa: The de facto rules', Paper presented at the Association of Collegiate Schools of Planning (ACSP)/Association of European Schools of Planning (AESOP), Joint International Congress, Toronto, Canada

Volbeda, S (1989) 'Housing and Survival Strategies of Women in Metropolitan Slum Areas in Brazil', *Habitat International*, Vol 13, No 3, pp157–171

Yapi-Diahou, A (1995) 'The Informal Housing Sector in Abidjan, Ivory Coast', *Environment and Urbanisation,* Vol 7, No 2, pp11–29

Chapter 12

Why the Urban Poor Cannot Secure Tenure: South African Tenure Policy under Pressure

Catherine Cross

For the poor in informal settlements, South Africa's urban tenure situation is turbulent, sometimes violent and generally insecure. High rates of demand for sites have contributed to high levels of informal urban land access processes. Informal land allocation and occupation deliver varying levels of security, and the major cities are struggling to bring the informal land process inside the orbit of formal planning and tenure. This attempt has proved to be slow and very difficult.

For the urban poor, many or most urban tenure systems, which are intended to give general security, either collapse back into informal or customary processes, or work very imperfectly. While some informal systems are equitable, others leave tenure subject to controlling power interests or to the elite, who may use the system as an income-generating activity and/or use control over the right to settle as a way to maintain power.

South African practice around securing urban tenure for the shack settlement population goes through the process of shack settlement upgrading, and is tied to the delivery of housing and services. Before the democratic elections of 1994, apartheid planning had a horror of an unstable, dissatisfied floating urban population that could be a potential force in civil disturbances (Dewar, 1992; Mabin, 1992), and tried to keep out rural in-migration. But since the lifting of influx control and the collapse of apartheid, political disturbances swept the urban sector and rural to urban migrants continued to arrive. South Africa's major cities started to make intense and almost desperate efforts to settle their arriving informal populations securely and comfortably.

This thrust assumes that the combination of housing and services is the key to installing the foundation for an escape from poverty. Here the urban planners' new efforts are likely to be well founded in principle. A secure home allows the opportunity for a household to invest with greater confidence and commitment (Agarwal, 1994). This promotes economic activity and the kind of upward movement from poverty that most households in South Africa's formally housed townships have achieved.

However, tenure security is treated as an afterthought, something that comes in as part of the housing package. In other words, tenure security is rarely, if ever, accessible other than through formal housing delivery. This process is a narrow gate, constrained by public budgets and capacity factors. The rate at which new households from the informal population are creating informal housing is overtaking housing and infrastructure provision. If this continues, many or most of the informal poor will remain indefinitely without secure tenure.

In the South African urban sector, tenure security means formalizing land rights through full formal private tenure. Urban private ownership, as the only fully recognized tenure in South Africa, became an end in itself for the poor. However, the route to formal security through private tenure is no easier for the poor in South Africa than in many other countries where land records are not as well maintained, and in some ways it may be more difficult. South African urban areas contain enormous pressures compared with other African countries. The obstacles that contribute to failures of registration and continuing informalization are: acute land shortage; conflicting tenure practices and land claims; politicizing land conflicts; high levels of competition over land; the race question; and high levels of civil and criminal violence.

Farvacque and McAuslan (1992) note that the requirements for effective urban land management include, among other factors:

- cheap and transparent registration processes with accessible records;
- flexible spatial planning and land use regulations;
- efficient land markets and limits on public intervention;
- systems that cope with landlord/tenant relations and can resolve disputes quickly;
- systems that offer a reasonable stake to leaders and gatekeepers for their cooperation;
- well trained and honest officials; and
- workable controls over transactions.

According to these standards, South African urban tenure for the poor would receive mixed reviews. The implicit objective is to transform the shack areas into formal low-income suburbs, known as 'townships'. Municipal spatial policies are taking shape around the discourse of corridors and nodes. In addition the land administration system is largely decentralized, with the large municipalities administering their own residents. But land administration is not free from central intervention in the form of land claim procedures. Disputed land claim processes can take years, stopping all development.

Houses in the formal townships can be bought and sold, though red-lining is reported to be a serious problem. For the informal areas there is very little that resembles a land market. Residents can sometimes sell their houses, or dismantle and remove them, but generally are unable to sell their sites. Registration is inaccessible to the informal population and there is no control over informal transactions. The relationship between formal land administration and community leadership is not structured and incentives in this sphere may be perverse. Informal landlord–tenant relations are unregulated. Whom to recognize as the landowner – men or women, a household alone or a household plus community – is a problem that has not been effectively dealt with in South Africa.

How urban tenure security works: The delivery link

Individual households in the shack areas are not able to gain tenure security. Most informal residents are occupying state or municipal land that is not for sale to individuals. If it were, poor informal occupiers would probably be able to obtain formal land security by paying for the survey and registration under the system used by high-income households. This would require them to show that their claimed site had identifiable boundaries, that they had signed an agreement with the previous owner and that the land was free of disputes with neighbours, as well as being free from any restitution claims, claims from the city, from developers, property owners or other parties. In settlements where the average household income is often less than R600 per month, very few could afford this route.

At present, the only practical way forward is a formal upgrading exercise, in which all the households in named informal communities are provided with registered sites along with infrastructure and services. However, the formal delivery process is itself complicated and expensive for public budgets. It also holds pitfalls for the poor.

Upgrading is carried out by the municipality or by authorized private developers. It usually entails arrangements with the owner to record the transfer of land en bloc for subdivision to the community members already on it. These delivery processes connect to national housing and tenure policy through the national (now provincial) housing subsidy, a one-off payment of R16,000 per household. This is intended to cover the nominal cost of the land and servicing, as well as the erection of the top structure. All upgrading work depends on developers being able to access the housing subsidy, paid over to them directly against their future delivery of serviced sites and often housing to a list of approved beneficiaries. These beneficiaries can access the subsidy only once. This process also grants the paperwork for private title when it delivers either a site with full housing or the site-and-service alternative.

Tenure security through housing also has a catch to it: obtaining and maintaining tenure means remaining in the serviced settlement. This entails being able to pay user charges for the services delivered, and there are likely to be hidden poverty thresholds involved. This is a high hurdle, even though recent policy moves may ease the problem.

For households from the informal areas, affordability is likely to limit access to tenure security even after upgrading and registration have taken place. This is most true in upgrading exercises where formal housing is part of the delivery undertaking, as it has been in Cape Town, but also holds true of site-and-service projects. At Blaauwberg in Cape Town's northern periphery, households that received serviced housing and tenure security moved out within two months. This was reportedly due to unsustainable and unanticipated new cost factors.

Movement from an informal area into formal housing is subjectively understood to imply a change in lifestyle: from frugal poverty self-identified as 'rural', in which households spend as little as possible, make do with substitutes where they can and rely heavily on networks and social capital, to a lifestyle with relatively expensive fixed standards of public consumption (Spiegel, 1999). Moving into a formal house or a township requires not only payment for services, but also a change over from what is perceived as a rural culture of mutualism and thrift, to an urban township lifestyle. This means buying formal furniture, appliances and other relatively expensive household goods, along with fashion clothing, food and other display items and consumables. Typical formal townships have household income levels two to five times higher than those of the typical shack settlement. This kind of cultural idiom also requires households to be much more autonomous in the economic sphere, and discourages rural style reliance on social investment and networks for insurance against shocks and reverses.

As Spiegel (1999) points out for Cape Town, most households living in informal areas do not have the income or resources needed to adopt conspicuous consumption. High urban consumption standards put a different light on cases of households moving out of newly upgraded settlements after only a few months because of affordability crises.

Planned municipal attempts to help the poor integrate into the urban fabric by subsidizing the upfront costs of the moves can be premature, and may result in the permanent loss of tenure security. This kind of well meant policy has the potential to backfire if the move carries other unanticipated or hidden costs, whether these costs are formal charges for services or social levies. Out-migration by the poor from upgraded settlements need not be due to any deliberate downmarket raiding by more advantaged households. It may merely reflect a predictable market process in which households that cannot afford township accommodation sell out to households that can.

This sequence cuts off the only route to tenure security for the upgraded household, because they will not be eligible for a housing subsidy again. The implication that the shift to formal housing is likely to be unaffordable to many or most families in the informal areas is serious, but the full implication appears to be that tenure security for the poor in urban informal settlements is likely to be a narrow gate. It will be further complicated by the post-apartheid demands of the poor on the political process and by the commitment of the new national elite to repair the damage done by apartheid and provide the urban poor with first class upgrading. Ways around the problem need to be found, which will take into account both the conditions around tenure security and the promises that were made to the poor at the close of apartheid.

FROM THE TOP: URBAN TENURE FROM THE NATIONAL LEVEL

National land policy and land-related development and tenure policy are formally under the department of land affairs (DLA). The new government created this department and the deeds registry has been located there since the democratic elections of 1994. South Africa's bitter history – of influx control, land confiscation and dispossession of formal and informal rights holders – meant that the DLA policy from 1994 to 1999 was driven by concern for giving the poor legal security and was directed to redressing the accumulated injustices of apartheid. Since the recent change of minister, priorities have become more specifically agricultural. In partial contrast, policies from the housing and services departments have consistently been more developmentally oriented. They have been more involved with planning goals and fast delivery of large land and housing schemes.

Exacting standards of participatory democracy in land matters have been required to meet demands from the non-governmental organizations (NGOs) that formerly opposed apartheid, as well as demands from the activists among the formerly disadvantaged. These processes have been necessarily slow in order to be seen to be both open and thorough. That is, delivery through the DLA has been subject to stringent public accountability and participation conditions, leading inevitably to lengthening queues. Brisk administrative action from the top down has been politically unacceptable in the post-apartheid climate, and budgets have remained underspent. However, delivery rates have been rising.

The DLA oversees tenure security and sets policy, but actual delivery is under the control of the metro and local authorities, transitional local councils, or tribal authority (TA) structures. At present the situation with regard to urban tenure in informal settlements is relatively fragmented, depending on what the underlying tenure is for the specific parcel of land. Choices for urban areas include: national government; provincial and municipal ownership; private sector corporate and private individual ownership; various forms of leasehold or delegated authority; different kinds of communal tenure under tribal or other institutions; and protected conservation status. A checkerboard of tenures and governing legislation has resulted in many urban areas.

Land administration and tenure administration are not centralized. Numerous functions and the administrative responsibility for land in certain categories have been devolved and delegated to the provinces. The three major functions of the DLA – land redistribution, land restitution and tenure – have been brought down from national control and placed under provincial and sometimes district offices. However, a range of other line departments are also involved in land and service delivery and are not formally linked to the DLA offices at any level. Under the complicated arrangement of delegated land powers that prevails, actual tenure control on the ground is sometimes weak and administrators who are frustrated, unsure of their authority and sometimes inclined to push the limits are handling land matters at many different levels.

In this grey area, fly-by-night developers and pirate land allocators in tribal and informal areas have been establishing room to operate. Their activities detract from tenure security for an unknown number of disadvantaged landholders with weak tenure standing or rights, which are difficult to document.

Larger-scale developments, which cross the boundaries of different kinds of tenure, are becoming characteristic in urban settings and are being held back by a lack of legislative framework. Some of these complications will be resolved by the current municipal demarcation initiative, which is adjusting the boundaries of designated cities and towns. But in other cases the addition of new areas will simply compound the problem of carrying out development across differing legal tenures.

THE CITIES: LOCAL PLANNING AND TENURE SECURITY

Durban in KwaZulu-Natal and Cape Town in the Western Cape are South Africa's major port cities; they carry the main share of economic activity that takes place in the coastal provinces. For both cities, the informally housed portion of the population is very large, reaching a third or more for Greater Durban but representing a smaller share in Cape Town. Both metro areas have experienced extensive land invasions as informal settlements have tried to occupy available undefended land. There are a number of areas in which the cities differ: the location and role of the former homelands as the main source areas for their informal populations; the existing tenures that informalization encounters; upgrading models in use; and the status of the rental market as a form of tenure.

The Western Cape has no former homelands anywhere in the province. Instead, it draws disadvantaged coloured in-migrants from the farms and old mission stations of the interior and disadvantaged black in-migrants from the neighbouring Eastern Cape, which contains the large, old and very poor former homelands, namely the Transkei and Ciskei. Accordingly, base tenure in Cape Town is relatively simple, comprising either private or government ownership at some level, with a substantial amount of barren land under conservation use. Government can acquire land for development or delivery purposes from the private ownership pool fairly easily, by the usual processes. Land claims are most often made around the extensive apartheid dispossessions of the urban coloured population, although black urban residents who were cleared from the Western Cape around the same time also enter the process.

Cape Town's informal settlers have moved almost entirely onto government and municipal land, which is likeliest to be undefended. However, Cape Town also maintains a relatively large rental market in housing, involving both private and public housing stock, which mainly serves the overcrowded coloured population but also accommodates black households. Secure formal leases are involved in an unknown but substantial proportion of cases, offering a strong degree of tenure security.

In contrast, Durban is the hub of a sub-regional system incorporating various pieces of the former KwaZulu homeland located adjacent to the city. Metro Durban has a range of legal tenures, including communal occupancy.

Informal land sales are common and accepted in the peri-urban zone of the TAs (Cross, 1992 and 1994).

In Durban's largest informal settlements, developed in the early 20th century on private black- and Indian-owned land, informal occupiers were not resisted because the land was run under site rental tenancy as a profit-making business and it was difficult for government to interfere. In the 1980s, when no new township housing was being built, policy encouraged informal infill in the townships, leading to crowding and social strain. Because most suitable land is already informally occupied, difficulties in acquiring land to develop or for delivery are mentioned repeatedly in the Durban spatial development plan (1998 and 1999). This plan calls for interventions to regularize tenure and acquisition processes so that they do not become indefinitely bogged down by conflicting tenures and unmanageable legal requirements.

Most informal settlements in Cape Town are concentrated in areas in and around the older townships and their railway and transport links and on marginal land near the airport. However, informal occupations are appearing more and more in established coloured and white neighbourhoods. No significant informal settlement has reached the central business district (CBD). Historically, economic activity has moved east and inland away from the present CBD, in the process overtaking the black and coloured townships planned for the periphery. Following this trend, informal settlements have moved away from the coast and the port into the new interior sub-centres on the eastern periphery, as well as appearing in patches to the north and south along the coast.

In Durban, a burst of intense activity in the early 1990s scattered informal occupancy throughout the central city in small colonies. However, the large areas of informal occupations remained peripheral, on private land 20 kilometres to the north at Inanda, the largest free-standing informal settlement in the country, and on township and TA land to the south of the port and industrial areas. These dense TA settlements are located both inside and outside the present metro boundary, which will move outwards under municipal demarcation planning.

Migration into both metropolitan areas has been rapid over the past 20 years. However, Durban's disadvantaged rural-origin population is long established and in-migration has probably peaked and is now slowing (Cross, 2000). After its clearances under apartheid, Cape Town has been receiving black in-migration from its source areas in the Eastern Cape only since the 1980s. In-migration is very fast and may be accelerating (Cross and Bekker with Eva, 1998). As a result, the informal areas in Cape Town are new, compared with Durban's, and are expanding. Formal housing delivery in both cities is behind the curve of demand, though the Cape Town shortfall is more serious (Abbott and Douglas, 1999; Cross, 2000).

Resources now flowing into the delivery process for the informally housed population are massive in relation to total metro expenditure (Durban Spatial Development Plan, 1998). Such flows may not be sustainable indefinitely. This cost to the cities of delivering living sites to the in-migrant rural poor raises more queries about changes to delivery models that will carry tenure security in the future, if alternatives are not found.

At local level in the metro urban areas, the picture has been further complicated by municipal efforts to provide a process that is fast, just, transparent, accountable, democratic, well targeted and cost-effective. Formal processes have been delayed in urban areas for a number of reasons: conflicting claims against apartheid land confiscations; a general desire to provide the poor with decent, first class tenure; and invasions of land and of housing by desperate groupings of the homeless. It is possible that the entire range of policy goals being attempted is partially contradictory.

The underlying informal tenure systems appear to be similar in Cape Town and Durban, as both are based on similar rural prototypes, adapted for use with faster urban turnover and less reliance on social capital assets within the local community. Results from Cape Town are mainly from Khayelitsha, a very large township that contains formal housing, informal settlements and open areas of unoccupied scrub (Cook, 1992; Cross and Bekker with Eva, 1999). Durban findings are from Mgaga, an upgraded informal settlement on former tribal land (Mbhele, 1999), from Inanda on former private land (Sutcliffe, 1985; Cross et al, 1992; Cross and Mbhele, 1994) and from Nyuswa, a peri-urban TA with very dense settlement and an informal land market.

Urban informal tenure systems are a thinner, fast action version of rural tenure. The rural model assumes that outsiders to the community join by approaching a local relative or contact. This person may provide land directly, or refer the candidate to a landholder who has a site that he or she is willing to dispose of. Obtaining land means getting citizenship, and only married couples with children are eligible. If an arrangement is reached, with or without direct payment, the sponsor takes the new entrant to the headman or chief, who interviews the candidate and approves the land transfer on behalf of the community, usually collecting a fee. The household of the new entrant stays with the sponsor while building a new house, in order for the neighbours to approve the prospective community members.

Up to seven levels of permission from the different principals may be necessary. Witnesses to the public site assignment secure tenure. Maintaining these tenure rights depends on good moral behaviour and observing the principles of mutual support, because crime or violent behaviour is supposed to mean expulsion. Latterly, chiefs have been behaving more arbitrarily, and may expel entrants.

The urban model thins down to one or two levels of approval. Survey results (Cross and Bekker with Eva, 1999) indicate that most black families who move into an informal settlement in Cape Town approach some form of local leadership through a sponsor, usually a network connection of the new entrant household, while some approach the leadership directly. Single women have least standing but normally get sites. Leaders who allocate land are the same individuals who are responsible for maintaining public order. A screening interview usually takes place, payments, if any, are made and a site is assigned immediately. The transaction should be recorded in the community books, but receipts are very rarely mentioned. The candidate rapidly takes up occupation and begins building before the site is reallocated. In a settlement with open land and little active leadership, some new entrants are able to move in without first seeking permission.

Maintaining tenure rights depends on being able to defend the household's position, as well as on good moral and social behaviour and not offending powerful local individuals. Both urban and rural tenure security for born outsiders is always contractual and depends on relationships between the landholding household and the rest of the community.

Rental tenancy as discussed here applies to Durban. It was common in all the informal areas up to the 1990s. Individual informal site renting contracts linked landlords and tenants. The tenants were responsible for putting up their own shelter, and for removing it when they left. Candidate tenants were usually introduced by contacts to the landlord, who would then screen the candidate. Evictions were relatively common if tenants were unable to pay their rent for a substantial period, and tenure security was perceived as variable. It was impossible for the city council to deliver services to these private tenure areas because each landowner had to agree separately. Landlords were unable to raise rents to recoup their costs and the tenants were not willing to pay more in order to obtain better infrastructure and services (Cross, 1994).

Durban's landlord–tenant relations exploded in the general violence of 1985, after the murder of Victoria Mxenge by security forces. Mainly Indian landowners were driven from the northern settlements (Sutcliffe, 1985; Cross et al, 1992). Over the following seven or eight years, the ANC-aligned youth movement supported a general boycott of rent payments as a civil disobedience strategy. Landlords as legal landowners were unable to return and re-establish their claim to control the land and collect rent. In effect, legal private ownership was overthrown in the Durban informal settlements. Because of the highly volatile situation, no level of the state would act to help landowners who were not white. Weak informal ownership claims emerged later out of occupancy, but by 1992 many occupiers said frankly that they did not know who owned the land.

By the 1990s local committees of comrades or residents administered the named settlement communities and informal leaders emerged. They maintained public order, screened prospective new entrants and gave permission to settle, initially without charge. The unenforceable titles were eventually bought up by a department of the former state, putting the land into legal public ownership and allowing the municipality to carry out service delivery. Rent tenancy has now re-emerged, sometimes based on ownership conveyed by the delivery process.

Both metropolitan areas are working with spatial development frameworks that emphasize the compaction of settlements in order to reduce the costs associated with outward sprawl, both to the household and to the municipality (Cape Metropolitan Council, 1996; Durban Metropolitan Council, 1998 and 1999). Compact city models are based on promoting development and settlement in areas with location advantages around transport corridors and activity nodes. The goal is to create a denser but more accessible city core while locating new settlements, as much as possible, to unoccupied or lightly occupied areas close to the core. But core periphery problems put into question delivery via these models.

Households would save on lower transport costs, the poor would be less isolated and there would be faster access to core amenities. For the municipality there would be saving costs on infrastructure because housing located on smaller plots would include costs of roads and new suburbs, as well as costs of existing transport services, and savings would probably be very substantial in terms of municipal budgets. There is concern that public spending on subsidized service delivery to informal settlements may only be sustainable if other infrastructure costs can be contained to make up the shortfall. New government policies to provide basic levels of water and electricity free of charge to all households will relieve some of the strain at a household level but will further burden public spending. Cape Metropolitan Area (CMA) planners are working to quantify the costs of extending the housing subsidy to all potential candidates. If the CMA cannot sustain its full current delivery targets, it will limit upgrading and cut back on delivery of tenure security.

For urban compaction to work, households in informal areas would have to support densification. As both spatial plans acknowledge, densification is not popular with the urban poor. The Durban plan of 1998 notes that densities of up to 50 households per hectare already occur in the informal settlements, in townships with infill and backyard shack developments. These densities are significantly higher than those found in white residential areas, reported as less than ten households, on average, per hectare, and also more than densities reported for black peri-urban settlements on TA land. Indian, coloured and black townships with less infill have densities of less than 20 households per hectare. In this light, the informal settlements are among the densest recorded settlements. To densify further, low-rise, multi-family dwelling units appear to be the only viable option, but are very unpopular and hard to manage.

From the viewpoint of the informal residents, very crowded settlements are problematic because they increase quarrels and violence, stress levels, health problems, crime, sanitation problems and spiritual attack or witchcraft. Not all the poor want to locate to the city core. Many reject core city crowding and social conditions and prefer the urban periphery. Travel costs hit the household budget in relatively small instalments and are easier for the poor to manage. Larger amounts needed to pay housing costs or service charges are difficult to save.

In the CMA, pockets of informal settlements appear to track new suburbs that push the urban edge. These settlements follow low quality jobs outwards as they are generated in new areas (Cross and Bekker with Eva, 1999). In Cape Town the majority of informal periphery residents walk to work locally and report no transport costs. If the informally housed poor need to locate within walking distance to informal work opportunities in order to have money left over from their characteristically low wages, it may challenge the cost assumptions associated with densification. Very dense core city areas are overtraded for low wage and casual unskilled jobs (Cross et al, 2000), therefore transport and access advantages of core locations are hard to demonstrate to many of the poor. And for poor families who do want a central location, available development land in the core zone is usually much too expensive for low-income housing that is not based on high-rise construction.

Serious questions may arise around tenure security achieved through the present subsidized housing development programme, if the intensive implementation of compact city planning based on showing the viability of further densification fails. Whether tenure security delivered via housing subsidies can maintain present levels of low-rise construction is an open question.

Experiences with subsidy-based delivery in Cape Town suggest that the kind of model based on rollover upgrading with full provision of finished housing on an adjacent site is not financially viable when backed by the subsidy at its present level. The model used in Durban, of site-and-service delivery only with in situ upgrading, appears to be more viable but may still be in question if beneficiary communities are unable to contribute. The viability of upgrading as a mechanism for delivering tenure security appears doubtful unless there is an additional top-up subsidy from the municipality.

Other factors raise further questions around tenure security through upgrading. In addition to direct and recurrent costs there are institutional problems and time demands involved with participatory development. Procedural restrictions and community expectations have been widely identified as causes for delays and cost overruns (Cross et al, 1995), and there appears to be some disillusionment with participatory planning models in both Cape Town and Durban.

At Macassar on the extreme south of the Cape Flats, a negotiated upgrading project reached the halfway mark and then collapsed when the developer ran out of money and folded his operation (the city appears to have refused to cover cost overruns). With the subsidy exhausted, the community is still waiting for delivery. At Bloekombos an upgrading, organized at the high tide of participation, lodged responsibility for funds with the local leader, who obtained as much as possible and then vanished, leaving the project stranded without a budget. The community has been waiting five years for delivery. Unemployed local workers evicted from their housing invaded Telkom-owned land at Marconi Beam. This occupation was upgraded successfully on a participatory basis by a public/private partnership (Saff, 1996). However, the group was very small and the time spent by the CMA housing staff in one-to-one community contact would have been prohibitive for a larger project.

In Durban, the results of participatory planning have been so uncertain that commercial developers often treat participation as a major liability for development projects (Cross et al, 1995). In addition to delays causing cost overruns, there have been problems around gatekeeping and illegal charges. Leaders are responsible for representing communities in the development process but have no formal position in delivery legislation and seem not to be legally accountable.

One development committee dealing with the city is reported to be charging high levies to community members, ostensibly as a contribution towards the cost of service delivery. In fact, the city has not charged the community any contribution and the money is widely believed to be going to the development committee leadership instead. Faced with an angry community when the fraud was discovered, the committee is said to have hired an enforcement squad, which killed several critics during night-time raids. Community members are now seriously frightened and it is reported that out-migration has begun. In the north, a woman community leader is reported to have taken over the role of landlord after starting out as a committee member. She has been resisting the delivery process because tenure security threatens her position and her source of revenue. Both kinds of event are thought to be fairly common (Ward and Chant, 1989, for international examples).

Emphasis on participatory processes may be contributing to local leaders taking advantage of their opaque and unaccountable positions to exploit their constituencies as an income source. Violence is reported over the right to control development in different settlements. From the delivery side, participatory processes can be so difficult to manage in their current form that the requirement of participation may be affecting delivery decisions negatively, as has been reported from Durban.

Some CMA development workers have come to see site allocation through bureaucratic processes as the most effective route to follow. There is less opportunity for disappointment and blame, which cause delays, than with an intensely personal participatory process. It is also clear that accountability is not well provided for and can sink projects. When planned upgrades or delivery initiatives do not go through to full completion, the community cannot take over and finish off informally; instead, the project hangs in limbo indefinitely. The lost subsidy funding cannot be restored and title is never delivered.

Conclusions

At present, informal settlement upgrading is the only process that can provide formal tenure security to the informally housed population. In the current delivery framework, this kind of security appears very problematic from the metro evidence. The rate of delivery is far behind the need and is inadequate. Rules and procedures are involved and rigid, cutting off alternatives to the current system. Even when full housing delivery is not a requirement, this tortuous route to tenure security may be unaffordable for the coastal metropolitan cities and unattainable for most of their informally housed residents.

It is not clear whether current planning models will be acceptable to residents of informal areas, or whether they will contain costs by a large enough margin to free up subsidized delivery and therefore release the constraints on access to tenure security. Participation models present problems around accountability and costs, and hidden social costs can make tenure delivery unsustainable even after it is formally completed. Urban informal tenure currently gives fair residential security but not the economic rights to support household-based investment, and allocation is subject to corruption.

Tenure security needs to break free of the upgrading process so that informal housing can be made legally secure. Work is currently being undertaken on low cost mechanisms in order to legally link housing to owners without going through the upgrading process. Examples are the Thaba'Nchu urban tenure project by the DLA and geographic positioning system (GPS) linked research by land surveyors in Cape Town and KwaZulu-Natal. These efforts need to be supported. However, for tenure security outside housing delivery to be successful, other issues need attention. These include:

- local leaderships who see their continuing interest in keeping full control of allocation and rejecting household tenure security;
- lack of accountability provisions for community representatives in negotiation processes;
- lack of complaint and adjudication mechanisms that are workable, well known and accessible;
- lack of effective access to registration and records;
- potentially opposed interests of women, men and children, particularly in unmarried households;
- possible detours from securing tenure into rental tenancy, which carries worse conditions than informal occupation itself; and
- viable physical housing models and a viable physical planning model to sustain tenure security.

A further serious concern is the potential downside of tenure security for the very poor. Land under private tenure is a dangerously slippery asset for people with inadequate income levels. The examples given of newly titled owners selling out when faced with hidden costs illustrate this. Title protects owners from outside attack, but not from poverty and distress sales. If private tenure is economically unsustainable for the household, the loss of new rights along with previous housing pushes the poor deeper into destitution.

It also needs to be remembered that property rights are a powerful determinant of the nature of society. Tenure security in what is now a partly unstable communal system of allocation and occupation will change the fundamental property rights that underpin informal communities. As yet there is no clarity in South Africa on family property law or on communal tenure systems. Depending on how rights are assigned, men, as assumed household heads, may be unable to kick out women and children any longer, but on the other hand families that are already precarious may break up, and communities may find themselves unable to require acceptable behaviour or expel criminals even when they desperately need to. There is a need to follow up the socio-economic impacts of tenure security before changing policy.

Acknowledgements

The author would like to thank Nontando Nolutshungu for help with the results from Cape Town and M T Nzama and O S Dlamini for help with the results from Durban.

References

Abbott, J and Douglas, (1998) *Informal Settlements in the CMA: Status and Trends Report, First Draft*, unpublished research report, Cape Town Metropolitan Council

Agarwal, B (1994) 'Gender and Command over Property: A Critical Gap in Economic Analysis and Policy in South Asia', *World Development*, Vol 22, No 10, pp1455–1478

Cape Metropolitan Council (1996) *Metropolitan Spatial Development Framework: A Guide for Spatial Development in the Cape Metropolitan Functional Region*, Regional Planning Technical Report, Cape Town, April

Cook, G (1992) 'Khayelitsha: New Settlement Forms in the Cape Peninsula', in D Smith (ed), *The Apartheid City and Beyond: Urbanization and Social Change in South Africa*, Routledge, London

Cross, C (2000) *The City as Destination: Migration Trends in the Durban Metropolitan Area, Based on the 1999 Quality of Life Survey*, unpublished report to Durban City Council

Cross, C (1994) 'Shack Tenure in Durban', in D Hindson and J McCarthy (eds), *Here to Stay: Informal Settlements In Kwazulu-Natal*, Indicator Press, Durban

Cross, C (1992) 'An Alternate Legality: The Property Rights Question in Relation to South African Land Reform', *South African Journal on Human Rights*, Vol 8, No 3

Cross, C and Bekker, S with Eva, G (1999) *En Waarheen Nou? Migration and Poverty in the Cape Town Metropolitan Area*, unpublished report, Cape Town City Council, Cape Town

Cross, C, Clark, C and Bekker, S (1995) *Coming Clean: Creating Transparency in Development Funding*, Indicator Press, University of Natal, Durban

Cross, C, Bekker, S, Clark, C and Richards, R (1992) *Moving On: Migration Streams Into and Out of Inanda*, Natal Town and Regional Planning Commission Supplementary Report, Vol 38, NTRPC, Pietermaritzburg

Cross, C, Mbhele, W, Masondo, P and Zulu, N (2000) *Employment Issues and Opportunities in the Informal Economy: Women Home-based Workers in Durban's Shacks and Townships*, unpublished report to International Labour Organization

Cross, C and Mbhele, W (1994) *Leadership and Patron–Client Relations in Inanda*, unpublished research report, Human Sciences Research Council, Pretoria

Dewar, D (1992) 'Urbanization and the South African City: A Manifesto for Change', in D Smith (ed), *The Apartheid City and Beyond: Urbanization and Social Change in South Africa*, Routledge, London

Durban Metropolitan Council (1998) *Durban Metropolitan Area Spatial Development Plan Vol 1, Spatial Development Framework*, Urban Strategy Department for DMA Spatial Development Steering Commitee, Durban

Durban Metropolitan Council (1999) *Durban Metropolitan Area Spatial Development Plan, Vol 2, Strategy and Action Plan*, Urban Strategy Department for DMA Spatial Development Steering Commitee, Durban

Farvacque, C, and McAuslan, P (1992) *Reforming Urban Land Policies and Institutions in Developing Countries*, Urban Management Program Policy Paper 5, World Bank, Washington, DC

Mabin, A (1992) 'Dispossession, Exploitation and Struggle: An Historical Overview of South African Urbanization', in D Smith (ed), *The Apartheid City and Beyond: Urbanization and Social Change in South Africa*, Routledge, London

Mbhele, W (1999) Field notes, unpublished

Saff, D (1996) 'Claiming a Space in a Changing South Africa: The "Squatters" of Marconi Beam, Cape Town', *Annals of the Association of American Geographers*, No 86, p2

Spiegel, A (1999) 'The Country in Town? Reconstitution of a Discarded Dichotomy?', paper presented to international conference, *Between Town and Country: Livelihoods, Settlement and Identity Formation in Sub-Saharan Africa*, Rhodes University, East London, June

Sutcliffe, M (1986) *Attitudes and Living Conditions in Inanda: The Context for Unrest?*, unpublished report, Built Environment Support Group, University of Natal, Durban

Ward, P and Chant, S (1987) 'Community Leadership and Self-help Housing', monograph, *Progress and Planning*, Vol 27, pp69–136

Chapter 13

Privatizing Displaced Urbanization in Greater Nelspruit

Cecile Ambert

Liberalization and privatization are held to be indispensable conditions of development. These trends in governance which assume that development materializes through the free market are impacting on legislation, policy and programmes in South Africa. The South African government is nominal owner of most of the land and housing located in the former homelands and self-governing territories, representing almost 13 per cent of the total South African territory. Most towns created to further apartheid's displaced urbanization policy in these territories are now targeted for tenure reform.

The neo-liberal rationale provides that the privatization of the land vested in the nominal ownership of the state will satisfy development needs at the local level. This chapter examines whether tenure reform can lead the way for the integration and development of displaced urban areas in a neo-liberal environment, with specific reference to the upgrading process under way in the Greater Nelspruit area.

Greater Nelspruit is situated in Mpumalanga, in the north-east of the country. It is home to an estimated 250,000 residents and is characterized by a remarkable diversity of settlement types. The area comprises two proclaimed and surveyed townships established under Proclamation R293, KaNyamzane and Matsulu (although Matsulu B is unsurveyed and under the de facto jurisdiction of a traditional authority), and four areas under traditional authority jurisdiction, as provided for by Proclamation R188. They are Zwelisha, Dantjie Pienaar, Luphisi and Msogwaba Mpakeni. KaNyamazane is closest to Nelspruit, approximately 20 kilometres to its east. Matsulu and Luphisi are 40 kilometres and 35 kilometres north-east of Nelspruit, respectively.

Proclamation R293 towns are characterized by nominal ownership vested in the office of the minister of land affairs, limited property rights granted to the residents and inadequate measures provided for surveying and registering rights.

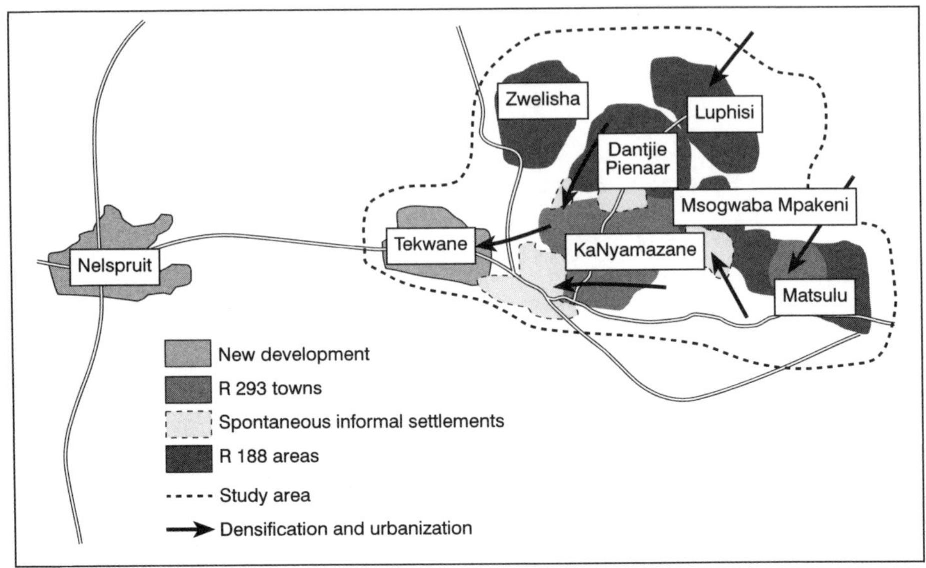

Figure 13.1 *The Greater Nelspruit Area*

Proclamation R188 areas are unproclaimed and seldom surveyed and the residents have no ownership rights.

The far eastern and northern settlements of Luphisi and parts of Dantjie Pienaar can be defined as 'closer settlements', dense settlements in the former homelands, far from centres of employment and commerce. The core of KaNyamazane, established in the 1960s, is a former homeland border town. The residents of this dense settlement commute daily to the nearest city or points of employment, in this case Nelspruit and white-owned farms (Centre for Development and Enterprise, 1998). Parts of Matsulu and KaNyamazane established at a later stage are peri-urban, traditional tenure, mixed settlement type with traditional dwellings and tenant and subtenant infill, whose inhabitants commute daily to places of employment (ibid). Illegal settlements located on state-owned land and on commercial farmland have mushroomed in the last 15 years. Tekwane, situated between Nelspruit and KaNyamazane, heralds the implementation of the government's housing policy in the form of an individual ownership-driven greenfield development.

The tenure types in the area are individual ownership rights, 'deed of grant' rights and 'permission to occupy' rights. Ownership rights confer permanent and exclusive rights to the use and possession of the property upon the individual. Few of the residents hold ownership rights. Most residents in the proclaimed areas have deed of grant rights that are said to compare favourably with ownership rights. However, the expropriation provisions are more arbitrary than for individual ownership, as they do not require the consent of the rights holder. In parts of Matsulu, where the number of residents exceeds the total number of sites, permission to occupy rights overlap with deed of grant rights. Permission to occupy rights pertain to the top structure and not the land, and can be dismissed summarily by the issuing authority. Most residents of the R188 areas hold permits to occupy, which are conditional on the payment of a nominal amount determined by the traditional authority. There is also a considerable amount (23 per cent) of rental, including subletting.

The tenure upgrading process, which is in accordance with the Department of Land Affairs (DLA) in situ land tenure reform approach, was initiated in 1998 by the local authority, in preparation for the extension of municipal services in the area. It is being financed and managed by the DLA. The other stakeholder involved, in terms of land allocation, is the traditional authority.

Government policy is to promote diverse forms of tenure including individual ownership, group ownership and rental. Nevertheless, individual ownership has been selected as the only tenure option for this project. The choice was motivated by perceptions about ownership as a prerequisite for the extension of municipal services and payment for them, ownership as a motivating factor for housing consolidation, and ownership as a catalyst for local economic development. Apart from the considerations of social redress, which generally underpin the drive towards ownership-driven land reform, the privatization of state-owned land in the area is also expected to be a catalyst for development.

However, during the participatory planning process residents were asked to rank various issues in order of priority. Ownership was only seen as a high priority in two of the six areas. These were both R188 areas, Dantjie Pienaar and Luphisi. Zwelisha residents did not select it as a priority issue at all and it was a low priority issue for residents of Matsulu B. It was also a low priority or not selected at all in the two R293 towns.

SERVICE DELIVERY AND PAYMENT FOR SERVICES

The perception that ownership is a prerequisite for the extension of municipal services and payment for services is widespread and relies primarily on the notion that ownership will foster civic pride among the residents and a greater propensity to pay service charges and rates. In addition, the local authority argues that the delivery and maintenance of infrastructure necessitates municipal ownership of the land on which bulk services are provided and the identification of end users for revenue generation (R Coetze, 1999, interview).

At present in R188 areas, most of the residents rely on standpipes and wash areas, and service charges are not levied. The Greater Nelspruit transitional local council (TLC) is now responsible for the servicing needs of the residents of the R293 towns and the R188 areas and minimum services are to be extended throughout.

Service levels are lowest in Matsulu B, where 60 per cent of residential stands have services below the minimum national service standards of a 'ventilated improved pit latrine' on each stand and shared standpipes at specified intervals. While this is the area with the fewest stands, it is also the area where there is most contest over land ownership. It was proclaimed as a formal township in 1992 but the proclamation was never put into effect, so that the de facto traditional authority of Chief Dlamini, who is vehemently opposed to the conversion of rights to ownership, is still enforced (Chief Dlamini, 1999, interview).

There is a significant diversity in income patterns in the study area. Income levels and residential costs are lowest in the R188 areas (D Thlabati, 1999, interview). Few of the residents of these areas, where the impact of the improved provision of services and the concomitant costs will be felt most, can afford to pay service charges. Even in the R293 towns, where many residents can afford to pay for services, the existing service payment levels stand at around 25 per cent.

This questions the sustainability of the delegated management initiative for the extension and upgrading of water services undertaken by the municipality. Should both the payment levels and the amount being paid for services not increase, the viability of the initiative is threatened, as the local authority would have to subsidize the end users heavily.

Ownership as a Motivator for Housing Consolidation

A number of stakeholders in the upgrading project were interviewed. They included DLA officials, local authority officials, local councillors and the mayor, consultants in charge of planning and conveyancing, and professionals involved in upgrading projects in other parts of South Africa. The majority of those interviewed mentioned that access to housing subsidies was conditional on the provision of individual ownership. However, the national housing policy aims to promote diverse tenure options and the housing subsidy scheme allows for institutional ownership.

Ownership was identified as a factor for consolidation of housing and as a catalyst for the emergence of a flourishing property market (K J Van Rensburg, 1999, interview). In KaNyamazane, Matsulu A and C and, to a lesser extent, in the informal extensions of KaNyamazane, individuals appear to have invested considerable amounts in consolidating and extending their dwellings. This evidence questions the rationale that the granting of ownership results in an increased propensity to invest in housing consolidation. Rather, it could be argued that factors such as the age of the settlement and the relative proximity of the settlement to centres of economic activity promote its desirability as a residential option.

Ownership as a Catalyst for Local Economic Development

The interviewees felt that economic development is conditional on ownership, which allows local entrepreneurs to access loans to grow their businesses and outside investors to safely establish a foothold in the area (I Khoza, 1999; K

Van Rensburg, 1999, interviews). It was also asserted that the tenure upgrading process would bring the informal economy into the formal realm. Access to ownership is expected to halt the practice of loan holders who do not have registered property taking out life insurance policies as collateral. If it is not necessary to buy insurance policies, that money is freed up to circulate in the local economy (K J Van Rensburg, 1999, interview).

The neo-liberal, macro-economic policy of the government – the growth, employment and redistribution strategy (GEAR) – has paved the way for the formulation of departmental policy which could have a negative impact on the viability of tenure upgrading processes in areas such as Greater Nelspruit. For example, the proposed decrease in public transport subsidies is likely to raise the overall commuting costs of the residents of KaNyamazane, and more so those of Matsulu, Mosgwaba Mpakeni and Dantjie Pienaar. This will diminish the residents' savings capacity. Similarly, the proposals to extend the property taxation system across the country will have a direct impact on the socio-economic conditions of the residents.

There is significant evidence that financial institutions are reluctant to engage with the low-income section of the population and have traditionally shunned this market. It is doubtful that the value of most of the residential units in the area would yield sufficient returns to entice their interest. Similarly, low-income profiles in a specific geographic context are unlikely to attract outside private investment in the formal sector.

It is also unclear how the regularization process will accommodate the more agrarian R188 areas, if uniformity of allocation is sought throughout the area. Defining the size of plots could result in a substantial decrease in the availability of land for economic purposes.

Tenure reform from an urbanization perspective

Various national government departments have a stake in tenure reform as it affects their activities and the viability of their policies and programmes. Equally their policies and activities affect the outcome of the upgrading process. Table 13.1 illustrates the impact that the policies and programmes of these departments have on the in situ upgrading approach.

In order to maximize post-upgrading development, the need to coordinate upgrading initiatives with other departments, whose programmes and policies have an impact on human settlement, should be recognized. Failure to take these policy approaches into consideration could result in contradictory developments at the local level.

Spatial impact

The confirmation of rights held in the former homelands and self-governing territories assumes an in situ approach, which runs the risk of cementing inherited inequitable and inefficient patterns of settlement. In the current

Table 13.1 *Legislation and Policy Impacting on In Situ Tenure Regularization*

Government department	*Nature of policy and programmes*	*Impact on in situ approach*
Department of Water Affairs	Extends services to where people reside Requires payment for services	Supports DLA's in situ approach Raises issues of affordability
Department of Constitutional Development	Demand-driven funding for bulk services supply Extends municipal property taxation throughout South Africa Requires payment for services	Requires sustainability of funding, could conflict with DLA's approach Raises questions of affordability of ownership and general issues of affordability
Department of Transport	Decreases overall subsidies for transport	Spatial alienation of residents of most peripheral settlements
Department of Housing	One-off, demand-driven subsidy for greenfield or in situ development	Depending on location could strengthen DLA's approach or make it redundant
Department of Trade and Industry	Identifies special areas for public investment	Positive impact if the peripheral settlement falls within the special areas for public investment
Provincial services (health and education)	Core services and redeployment approach but limited budget	Could benefit residents of peripheral settlements, but highly dependent on budget
Municipal government	Extends municipal services Core services approach	Supports in situ tenure upgrading Could raise issues of affordability and municipal financial sustainability

macro-economic context, this approach could fail to adequately address complex socio-economic and political needs.

The adoption of a single ownership-driven approach to a local area marked by diversity tends to negate the policy of promoting diverse forms of tenure. Worse, the lack of coordinated action observable to date about where public investment is, and should be, occurring, could entrench the patterns of unequal access created by apartheid policy and legislation.

At present, the national sphere of government is divesting itself of its developmental responsibilities by: privatizing state owned land; promoting delegated management; formulating delivery policies which limit its contribution to one-off subsidies; curtailing the scope and nature of its services; and seeking to broaden the generation of revenue for the state. This results in a paucity of avenues for sustainable development in peripheral, underdeveloped areas like the ones described in this chapter.

Feasibility

If transferring nominally held state land to its real owners entails in situ regularization, to what extent can and should other departments, including local authorities, be prioritizing investment programmes in these areas? It is evident that a large proportion of households will benefit from the upgrading process. However, some might not be able to sustain the additional burdens brought about by ownership without significant additional investment from a number of government parties. Given the multiplicity of contexts of developmental need in South Africa and the limited financial resources available, unpopular policy trade-offs may have to be made regarding where and how public and private investment is located. Achieving both social redress and long-term economic sustainability may prove to be mutually incompatible goals. However, unless a strategic development approach that takes into consideration the contextual diversity of underdevelopment is devised and implemented, the lack of coherence and integration of government action will thwart development at the local level.

THE CHOICE OF OWNERSHIP AS THE PREFERRED TENURE OPTION

The Greater Nelspruit upgrade area is characterized by diversity in service levels, density levels, housing types and socio-economic profiles. This suggests that adopting a uniform approach to tenure upgrading could be problematic.

The vesting of ownership rights in the individual brings with it substantial costs including local property taxation, municipal levies and conveyancing costs. These costs are seldom communicated to the future property owners during an upgrading process. Privatization of land in such a context could result in the privatization of financial liabilities rather than the strengthening of tenure rights at the local level.

The onerous procedural obligations of legal property transfer could, as they have elsewhere in South Africa, lead to the perpetuation of informal land transfers and the renewed corruption and redundancy of the land registration system. The substantial fiscal expenditure incurred to upgrade tenure rights could become futile.

Project level education for the intended beneficiaries of tenure reform should target ownership and other forms of tenure, the provision of services and payment for services and municipal rates. The education strategy should seek to empower rights holders with information to make informed decisions by imparting an understanding of their current situation, the tenure delivery options available to them and the impact that each tenure option will have on their current situation.

The choice of ownership as the preferred tenure option will have varying impacts on different interest groups. For example, subletters and sharers may be excluded. Attention must be given to the release of additional or alternative land, where the tenure upgrading process could result in the displacement of existing residents. The role of municipal planning in the conversion process should thus be emphasized.

The identification of stakeholders in a particular settlement, and clarity about their contributions and interests, also emerges as an important procedural consideration. In the Greater Nelspruit context, significant tensions have been experienced between various stakeholders. It is likely that competition will also emerge at the settlement level between and within households. The Nelspruit study reveals that where land hunger is great and competition for the power to allocate land is stiff, the potential for violence and conflict is heightened. In Matsulu B, where Chief Dlamini has de facto power over the allocation of land, his opposition to the conversion of rights to ownership could potentially lead to conflict arising during the upgrading process.

Alternatives to ownership

The objectives of the new legislative framework in South Africa appear to be to confirm and protect existing de facto rights and to create a mechanism whereby these rights can be registered. It is important to note that some of the functions that individual ownership fulfils are also met by other tenure types such as group ownership. In formulating what is the most appropriate form of ownership for a particular person, family and/or community, one needs to understand which of the benefits accrued from ownership are most important for that person, family and/or community. Individual ownership is costly and should be considered as the secure tenure route only when ownership of the property is needed to provide other benefits, in addition to providing security of tenure.

In contexts similar to that of the Greater Nelspruit area, some households will benefit from a facilitated access to ownership, while the needs of others will not be adequately addressed by such a process. In order to accommodate the latter, use should be made of other existing forms of tenure, namely rental and group ownership. In addition, a new statutory form of tenure should be made available as one of the choices in a locally driven tenure upgrading process. This should provide burden-free secure tenure, where rights holders are protected against deprivation, have decision-making power in respect of the land and have the right to benefits accruing from the land.

Acknowledgements

This chapter is based on research conducted by the author at Development Works, an urban development organization based in Johannesburg, South Africa.

References

Centre for Development and Enterprise (1998) 'South Africa's "Discarded People": Survival, Adaption and Current Challenges', *CDE Research Series,* Johannesburg

List of Interviewees

Chirwa, P (Earthspace, Nelspruit)
Coetze, R (Nelspruit TLC)
De Villiers, N (Van Rensburg and de Villiers, Pretoria)
Chief Dlamini (Nkosi, Matsulu B)
Geldenhuys, N (Deeds Registry, Pretoria)
Hadebe, H (DLA Mpumalanga, Nelspruit)
Khoza, I (Nelspruit TLC, Mayor, Nelspruit)
Mabuza, F (Nespruit TLC, Deputy Mayor, Nelspruit)
Maseka, B (Nelspruit TLC, Councillor, Nelspruit)
Mazibuko, P (Nelspruit TLC, Councillor, Nelspruit)
Oosthuizen, S (Nelspruit TLC, Urbanization Department, Nelspruit)
Pirie, A (Nelspruit TLC, Councillor, Nelspruit)
Rademeyer, J (Rademeyer and Van der Merwe, Nelspruit)
Steyn, B (Nelspruit TLC, Urbanization Department, Nelspruit)
Thlabati, D (DLA Lowveld Regional Office, Nelspruit)
Van der Merwe, C (Africon, Pretoria)
Van der Merwe, L (Africon, Nelspruit, telephone interview)
Van Rensburg, K (Van Rensburg and de Villiers, Pretoria)
Van Rensburg, K J (Earthspace, Nelspruit)
Van Rensburg, M (Planpractice, Nelspruit, telephone interview)
Van Wyk, E (Van Wyk Conveyancers, Nelspruit)
Weich, G (Africon, Pretoria)
Williams, C (DLA)

Chapter 14

A Land Management Approach for Informal Settlement in South Africa[1]

Colin Davies and Clarissa Fourie

Although present government policy is to provide housing, services and security of tenure to people living in informal settlements, the United Nations Centre for Human Settlements (UNCHS) (1996, p292) estimates that 20 to 80 per cent of urban growth in developing countries is informal. Therefore, it is unlikely that the formal land delivery system in South Africa will ever catch up with the demand for formal housing. In the interim, this demand is being met by the informal land delivery system. In other words, informal settlements are likely to continue to provide shelter for many poor people who are waiting for a house within a formal development. Therefore, local authorities should develop mechanisms to manage existing and future informal settlements. This chapter will describe a land management approach for informal settlements from a local authority perspective. A number of themes are explored. They include:

- the role of local authorities in land management and informal settlement upgrading;
- the definition of informal settlements for the purposes of local level land management;
- a social change approach, used to understand and predict changes in informal settlement land tenure, based on Fourie (1993);
- the design of a land record using land information system (LIS) technology, based on an informal settlement in East London; and
- some important aspects for future research.

1 This chapter was originally published as an article in 1998 in *The South African Journal of Surveying and Mapping*, Vol 24, Nos 5 and 6, Colin Davies (ed), August and December

The role of local authorities in land management

In South Africa government is 'constituted as national, provincial and local spheres ... which are distinctive, interdependent and interrelated' (Constitution of the Republic of South Africa, section 40(1)). A municipality is the local level, or sphere, of government and is generally referred to as the local authority. The constitutional objectives of local authorities are to:

- provide democratic and accountable government for local communities;
- ensure the provision of services to communities in a sustainable manner;
- promote social and economic development;
- promote a safe and healthy environment; and
- encourage the involvement of communities and community organizations in the matters of local government (ibid, section 152(1)).

These objectives reflect the national government's intention (white paper on reconstruction and development) to 'bring the government closer to the people' (Parliament of the Republic of South Africa, 1994, p18). Local authorities therefore have a responsibility to manage land in a way that facilitates holistic development with the resources they have. Accurate, current and accessible land record data are essential if a local authority is to carry out these functions effectively.

Durand-Lasserve (1996, p70) recognizes that local authorities have 'a central role to play in ... the launch and implementation of an integration programme for irregular settlements'. In undertaking its responsibilities, the East London municipality is actively involved in upgrading and regularizing informal settlements within existing regulatory frameworks. Community participation is facilitated through public meetings and regular meetings between the community and the local authority. However, Durand-Lasserve (ibid, p70) also argues that 'local authorities seem to be in the best position to develop other ideas and methods for land management'. One of these methods is to develop a land management approach based on a partnership between the local authority and the settlement. The partnership approach recognizes that,

> *the settlement would indeed appear to be the most appropriate level for assuring an effective continuity of control over urban land use ... (and for assuring) ... the operation of urban services in relation to cost recovery* (ibid, p73).

This chapter examines such a land management approach.

Expanding the definition of informal settlements

Informal settlements are a global phenomenon and occur mostly, but not exclusively, in developing countries (Payne, 1989, p1; Fourie, 1993, p4; Durand-Lasserve, 1996, p1). The literature shows that the terms 'squatting' and 'informal settlement' are used interchangeably (Peil and Sada, 1984; Urban Foundation, 1991; Harrison, 1992; Fourie, 1993). The term 'informal' is used in preference to 'squatter' because of the negative connotations associated with the latter in South Africa.

A large percentage of urban residents in many developing countries live in irregular or informal settlements. The United Nations (Durand-Lasserve, 1996, pxviii) suggests that this figure could be as high as 70 per cent, with informal settlements accounting for up to 80 per cent of urban growth. South Africa is no exception; informal settlements have become a significant part of South African cities over the past decade. The UNCHS urban management programme states that there is not always a clear divide between legal development and 'development taking place within the framework of informal law' (1991, p7). As Doebele argues,

> *current research, rather than depicting a duality of 'legal' and 'not legal', delineates a real world that contains very complex mixtures of formal and informal systems with infinite variations in between* (1994, p48).

The formal, or statutory, and local informal systems work together and are modified to suit conditions in specific local situations. Thus informal settlements should be defined in terms of a continuum model.

From observations made in the Mzamomhle informal settlement in East London, Davies has shown that the continuum approach holds for the levels of physical development, socio-economic profile and legality of tenure (Davies, 1998, Chapter 6). Informal settlements exist on a continuum between completely illegal and completely legal types of settlement. Further, in terms of a social change approach (see below), the position of a settlement on this continuum changes over time. Classical definitions of informal settlements are not broad enough to provide a base on which to develop local level land management policies for informal settlements. Instead, for the purposes of land management in South Africa, an informal settlement may be 'any settlement that exists on a continuum of development, and is jointly managed by the local community and the local authority' (Davies, 1998, p104). In addition, the local authority may provide certain services to the settlement, which are commensurate with the degree of legality of the settlement and the level of responsibility that it has regarding the settlement.

Understanding informal tenure using a social change approach

The overview of the social change approach has been adapted from Davies and Fourie (1996) and Fourie (1993). The approach describes the transformation of indigenous land tenure over time in an informal settlement. It explains the role of land tenure rules, transactional behaviour and the influence of external factors, over time, in a tenure system of an informal settlement and is based on the concept of two or more interdependent yet opposing cultures within an encompassing context.

In East London, land tenure in informal settlements is derived from rural tenure rules that have been modified for conditions within urban settlements (Davies, 1998, Chapter 5). Some of the characteristics of urban informal settlement tenure are:

- Newcomers must be sponsored and undergo a period of probation before being allocated land;.
- Residents must adhere to accepted community standards of land tenure-related behaviour.
- Vertical and horizontal power structures develop and compete with each other for access to land, power and resources.
- Individual land rights are subject to the community over right.

The social change approach is based on the dialectical relationships, or structural tensions, that exist within Nguni society (the Nguni polity is an ethnic group which occupies the areas of KwaZulu-Natal and the Eastern Cape). There are two dialectics associated with a group. These are referred to as the internal and external dialectics. The internal dialectic is the structural tension between groups within the community. The external dialectic is the tension between the internal dialectic and external groups, which is affected by a number of factors. These factors include urbanization patterns, local authority policy and other interventions that affect the local system or community. The internal dialectic consists of opposing yet dependent tendencies within a group, manifested by the tendencies towards fission and integration. Fission is the process of individualization, whereas integration is the tendency to form coalitions to strengthen group identity. This internal dialectic explains the tension between individual and group rights with respect to land issues within informal settlements.

The change approach goes beyond the conventional assumptions that a simple hierarchical power structure determines social behaviour within a traditional Nguni polity or group. Rather, it assumes that, within a group, a range of different power levels exist, with competing subgroups within each level. These subgroups compete horizontally and vertically for control of resources, access to land and political power. The importance of land tenure rules for the functioning of the group is recognized. However, these rules are not static but are manipulated by the subgroups as they compete with each other. In other words, the rules transform over time. As Fourie notes, 'indigenous systems of land tenure within urban areas in Africa adapt to urbanisation via a fluid social field, where land tenure rules are manipulated within a complex of social forces', and 'competition and negotiation for personal advantage around land rights involves coalition formation and entrepreneurship' (1993, p438). This behaviour directly relates to informal settlement development over time.

The manipulation of land tenure rules to acquire resources, land and power can be analysed using transactional behaviour approaches (Fourie, 1993). However, such transactional behaviour is subject to the approval of the community, so coalitions have to be formed by those wishing to introduce or oppose change. Groups have to trade-off against each other in an attempt to gain leverage to accomplish their aims (ibid).

The influence of external factors on a community is also responsible for changes in social behaviour and the local land tenure system. These changes are caused by an alteration in the existing structural tension between subgroups and power levels within the local community. That is, external factors that influence behaviour within an informal settlement will also influence the land tenure system. In urban areas the local authority is responsible for implementing upgrade projects initiated by the community through the Reconstruction and Development Programme (RDP). Although these are positive actions, they do generally cause unplanned changes at the local level because the structural tensions in the local system have not been accommodated in land management policies to date, and the local authority's interventions are manipulated by people in the local community to their own advantage. An awareness and understanding, by land managers, of such behaviour is becoming increasingly relevant for the sustainable regularization and development of informal settlements.

Benefits of the change approach for land management

The advantage of using the social change approach is that it provides the conceptual framework to:

- identify the main stakeholders regarding land in the local system and their historical and political relationships to each other;
- predict changes in the local tenure system based on the interaction between internal subgroups and external factors – special attention should be given to local authority and large development interventions;
- identify the geographical and relational areas where transactional behaviour and entrepreneurship in land, resulting in changes to the local tenure system, could occur;
- identify factors that could cause formal tenure to revert to an informal form of tenure over time;
- monitor changes in settlement land allocation procedures over time;
- ensure that local authority policies and rules remain relevant and applicable in an informal settlement; and
- link the local tenure system to a multipurpose land record.

The above framework enables a land manager to develop flexible and appropriate policies to guide the upgrading and development of informal settlements. When local institutional structures, internal relationships and external factors have been identified, the process of developing an appropriate and sustainable land management approach for informal settlements can begin. This approach could assist the local authority in taking the necessary steps to ensure that the process is not stalled or derailed by specific groups within the settlement.

A NEW LAND MANAGEMENT APPROACH FOR INFORMAL SETTLEMENTS

Land management fundamentals

Dale and McLaughlin note that 'land management entails decision-making and the implementation of decisions about land' (1988, p4). The International Federation of Surveyors (FIG) recognizes the contribution of land management to sustainable development by saying that land management is 'the process of managing the use and development of land resources in a sustainable way' (1995, p1). O'Riordan (1971) introduces the importance of the various dynamic frameworks which exist and influence land management decisions, especially the social and political aspects of land. He states that land management is:

> *the process of decision making whereby resources are allocated over space and time according to the aspirations and desires of [people] within the framework of [their] technological inventiveness, [their] political and social institutions, and [their] legal and administrative arrangements* (quoted in Nichols, 1993, p35).

From a more socially aware perspective, Okpala says that

> *land management ... involves a process of inventorying and recording land, its characteristics and changes thereon over time, policies for its development and use, and of apportioning it to appropriate uses on the basis of priorities determined in the society's interest* (1992, p262).

These definitions recognize the element of change in land characteristics and social priorities for land use. Land management policies and procedures that are appropriate and flexible should, therefore, form part of the development and growth strategy for an urban area.

Yet Williamson (1991, p45) notes that current research points to the deterioration of the urban environment in most developing countries. As Okpala states, without an adequate 'identification and inventory of land, efficient and effective land management is hardly possible' (1992, pp262–263). Therefore, land management policies for developing urban areas should ensure that all land is appropriately identified and inventoried. This includes land parcels in informal settlements, which are not currently included in the cadastre.

The role of local authorities in informal settlement management

A land management approach for informal settlements requires that the local authority recognizes the de facto (derived) tenure rights of the community. Further, the community's right to exist must be guaranteed subject to certain conditions. This recognition would be given by the local authority and would not imply or guarantee that the community would be granted de jure (real) tenure rights where the settlement is currently situated.

To ensure flexibility, the local authority should recognize that local communities are 'capable of planning their life, including the natural space where they live' (Rene, 1996). Official local authority policy should 'establish a space for participation' (ibid) so that affected communities are able to actively contribute to land management decisions. The participation process should facilitate dialogue between the settlement leaders who have extensive local knowledge, on the one hand, and technical experts on development and management issues, on the other. Cross supports this argument by saying that 'planners need to move beyond an analysis of legal tenure in their areas of operation', and that they should endeavour to understand the 'implications of perceived and experienced tenure on the ground' (Cross et al, 1992, p44). That is, professionals should recognize the importance of local knowledge and participation in determining appropriate land tenure solutions when regularizing informal settlements. Williams (1997, p15) notes that consultation is often passed off as genuine participation. In other words, development plans and proposals are prepared in advance by technical experts and approval is then sought from the community through a consultative process. He argues that 'community organizations rarely participate in the formulation of development plans', leading to a lack of 'substantive participation' (ibid).

A participatory approach to land management should enable the local authority to develop and implement unique, but appropriate and sustainable, solutions within a particular settlement. The adopted approach should not impose foreign land management procedures based on existing cadastral systems, but should allow the local procedures to remain and transform over time. The land record procedures should be based on existing community land management processes for the collection and maintenance of land record data. The data would be used both by the community and local authority for ongoing development and regularization projects. The approach would recognize and accommodate the differences between formal and community land management approaches, the former regulated by statute and policy and the latter derived from social, political and economic factors within the community. The approach should encourage people to record all land transactions and allocations, even if they contravene existing legislation. The land manager should ensure that information flows efficiently between the community and local authority to inform the land development and service delivery process. To explore this idea further, a proposal for a land record is outlined below.

A land record for informal settlements at local level

The importance of a participatory approach is that it builds a partnership between the community and the local authority for the creation and maintenance of a record of informal land allocations and transfers. A method for implementing this is to establish a local level land record based on a LIS. Local level land record management is supported by UNCHS (Habitat) as a solution to regularizing informal settlements and increasing tenure security (Durand-

Lasserve, 1996, pxii; UNCHS, 1996). This approach is becoming increasingly popular (Fourie and van Gysen, 1996; Jackson, 1996; Ezigbalike, 1996). It accords with current South African government policy where the 'intention is for a decentralized land administration (management) and land reform one-stop service that is largely based at third tier government' (South African Government, 1996, pvii). A local land record office would allow local tenure patterns to evolve while simultaneously providing a more formal level of security of tenure to the residents in the community.

Professional and local skills

A professional land manager, acceptable to the community, should be appointed by the local authority to manage the technical, operational and social components of the system. A local resident should be trained and employed as a local land administrator to work under the supervision of the land manager (Jackson, 1996, p283; Fourie, 1996; Landman, 1996). Their tasks would include basic surveying of new allocations, administration of the record system and assisting residents with queries or complaints regarding development and services. The land administrator would, therefore, provide a range of services to the local community. The community leadership should be integrally involved in the determination of what data should be included in the system. They should be empowered to use the data in participative decision-making on projects or developments that directly affect the settlement (Rene, 1996). This form of decentralized information gathering and dissemination could have a major impact on the way in which local authorities deal with informal settlements in the future.

Formalizing a partnership with the local authority

Negotiations between the community and the local authority should cover all aspects of why the land record is required. An agreement, with a non-cancellation clause, should be signed by all stakeholders (Cabannes, 1997, p13). Cabannes argues that a clause that legally binds the parties to the agreement, and specifies penalties should a party withdraw from the agreement, can be a key factor in ensuring the ongoing existence of such a partnership. The partnership agreement should specify and provide for:

- the parties to the agreement;
- the purpose of the agreement;
- the rights and responsibilities of each party;
- ownership and custodianship of the data;
- dispute resolution mechanisms; and
- penalties for withdrawal from the agreement.

The agreement should also clearly and simply set out the 'objectives and results which could be realized within a short period' and should specify 'the rights and duties of each party involved' (ibid, p4). In other words, the community, as a whole, should know how they are able to participate in, and benefit from, such a project.

Mapping of spatial relationships

The settlement should be mapped using a suitable method that will facilitate the creation of a LIS. A combination of geographic positioning system (GPS), terrestrial survey and digital mapping methods could be used to collect the spatial data required for the LIS. Digital mapping software could be used to geo-reference, mosaic and model non-metric aerial photographs taken from a light aircraft. Vectors could be traced over the image and exported into the LIS. A paper map could be plotted on an inkjet type plotter for use by the local land administrators. The following information could be included in the spatial data set of the land record (Davies, 1998, p118):

- service corridors;
- access routes;
- recreation areas and community services;
- utility positions (stand pipes, poles, drains, access covers);
- public facilities (public telephones, post boxes);
- proposed and existing land use zoning;
- level of services per land unit within the settlement;
- 1:50 year flood line – this is important when a river or stream runs through the settlement to prevent settlement below the flood line;
- geo-technical data, including water table levels and underground rock types; and
- topography, including a digital terrain model if available.

Mapping of social relationships

A questionnaire should be drawn up by the community, under the guidance of the land manager, to obtain social data for each household. Considering work in East London, the following should be considered for inclusion in the questionnaire (Davies, 1998, p119). However, user needs and local circumstances should determine the final format.

- date of enumeration;
- shack number;
- details of person who allocated the land;
- name of household head – family and first names;
- identity number;
- sex;
- marital status – single, married, divorced or living with partner;
- name of spouse or partner;
- identity number of spouse or partner;
- postal address, if available;

- combined income of household (this should only be asked if the information is to be used for a housing subsidy application, otherwise people may be reluctant to provide other information);
- number of children in the household, and their ages;
- date of arrival in the settlement;
- date of site allocation (seldom the same as arrival);
- available services to the dwelling, if any;
- name and address of next of kin, for burial purposes;
- number of permanent and weekly lodgers in the household;
- whether the site is physically demarcated by fence or hedge;
- a physical description of dwelling;
- service account number and record of payments, if applicable; and
- record of previous transfers of dwelling, if any.

The data from the completed questionnaires should be entered into a database and linked to the spatial map of the area in a LIS.

Local records administration

To encourage trust and participation in the process, people should be given proof of their inclusion in the database. A pre-numbered certificate issued by the local authority should be completed for each person shown on the land record. This certificate should indicate the date, number of the shack, and name and identity number of the person reflected in the land record. This should be given to the owner, who should sign for it. The certificate should also be signed by the chairpersons of the executive committee and/or the area committee. A copy should be kept on file in the settlement. New certificates should be issued to newcomers, or to reflect changes in ownership of existing shacks. If an old certificate exists, it should be returned for filing in the record office before a new one is issued. The person disposing of their claim to ownership of the shack and occupation of the land should accompany the person acquiring the existing shack to the relevant community leader. The previous owner should endorse the old certificate to relinquish his or her claim. This procedure formalizes existing informal processes. A paper map, aerial image of the settlement and a tabulated record of shack owners should be supplied by the land manager to the local person administering the area. The local land administrator should be trained to annotate any new allocations on the map and photograph. Any land transactions, including new allocations, should be entered into the local filing system.

The local leadership should be held responsible for ensuring that newcomers to the area are incorporated into the system and that the local land administrator maintains the records to the satisfaction of the land manager. This gives them a sense of ownership of the system and should not undermine or threaten their existing power base. In addition, such a system would build capacity among the local leaders to make land-related decisions (Fourie and van Gysen, 1996, pp358–359) and give them an understanding of formal land management procedures. This is not the usual way in which professionals or local authorities deal with informal settlements at present.

Finally, at regular intervals, the land manager should visit the land record office to collect the annotated maps, image and updated textual records. These new data would be used to update the LIS. At the same time new paper copies of the maps and records should be left with the local person. A close check on issued and returned certificates should be kept, to reduce fraud. Any problems could also be discussed, guidance given, general information transferred and participatory management increased. This would encourage the long-term sustainability of the land record and registration system.

CONCLUSION

Current United Nations thinking argues that an important option for land management is to legalize 'the practices and systems which are currently not covered by or are excluded from legal structures' (Durand-Lasserve, 1996, p62). Therefore, a participatory approach for collecting, maintaining and using land record data for the regularization and development of informal settlements is proposed.

Informal tenure systems exist and change over time in response to changes in the social structural tension. These changes can be described by a social change approach. Current, accurate and accessible information is required for effective land management, but current land management procedures, based on the existing cadastral and registration systems, do not extend to informal settlements. As a result, development of these areas is hindered. The land management approach is based on a land record that uses a LIS to link spatial and social data for planning and administrative purposes.

The proposed land management approach will hopefully contribute towards the following objectives:

- Build land management capacity within local communities by introducing people to a mechanism for recording land allocations and transfers; community leaders should be trained and equipped so that they can encourage the community to participate in the land record.
- Protect the interests of the poor, weak and women (UNCHS, 1996, p291).
- Provide security of tenure to groups who may experience difficulty in accessing resources when settlements are upgraded or regularized. The data in the land record should provide evidence, over time, of land allocations to household heads, tenancy patterns, income profiles and sex of the household head.
- Strengthen the position of the local community during negotiation, planning and implementation of development and upgrade projects in terms of current policy. Therefore, the land record office should make land-related information available to the community and strengthen people's understanding of how information is used for planning and administrative purposes.

- Give residents more security of tenure, and provide a resource base to the local authority for the service planning that normally accompanies the regularization of tenure.

This approach would facilitate rapid delivery during an upgrade project and would help local authorities in developing countries with budgetary pressures.

Further research on the applicability of the social change and land management approaches in other African countries should be undertaken. Additional research could also focus on refining some of the concepts discussed above, with the objective of applying the advantages of modern technology to local conditions.

The plight of people living in informal settlements has highlighted the need to implement a land management approach that empowers people, while providing the necessary information for the planning and regularization of the settlement by the local authority. South African land surveyors should be encouraged to develop their skills, especially with regard to social and land management tools. They should actively assist local authorities and communities in developing solutions to the complex issues surrounding informal settlements, development and local authority land management in general.

REFERENCES

Cabannes, Y (1997) 'How and Why Does a Partnership Work?', *Habitat Debate,* UNCHS, March 1997, Vol 3, No 1, pp13–14

Cross, C, Bekker, S and Clark C (1992) 'People on the Move: Migration Streams in the DFR', *Indicator SA,* Vol 9, No 3, pp41–44

Dale, P F and McLaughlin, J D (1988) *Land Information Management: An Introduction with Special Reference to Cadastral Problems in Third World Countries,* Clarendon Press, Oxford

Davies, C J (1998) 'Land Management of an Informal Settlement in East London', Unpublished MSc Thesis, Dept Surveying and Mapping, University of Natal, South Africa

Davies, C J and Fourie, C D (1996) 'Incorporating a Social Change Model into Land Record Design', Paper presented at *International Conference on Land Tenure and Administration,* Orlando, Florida

Doebele, W A (1994) 'Urban Land and Macro Economic Development: Moving from "Access for the Poor" to Urban Productivity', in G Jones and P M Ward (eds) *Methodology for Land and Housing Market Analysis,* University College Press, London

Durand-Lasserve, A, assisted by Valerie, C (1996) *Regularization and Integration of Irregular Settlements: Lessons from Experience,* Urban Management Programme, UNDP/UNCHS/World Bank-UMP, Nairobi

Ezigbalike, I C (1996) 'Integrated "Cadastre" for Rural Africa', *South African Journal of Surveying and Mapping,* Vol 23, No 6, pp 345–352

FIG (1995) *The FIG Statement on the Cadastre,* The International Federation of Surveyors, Publication 11, Canberra, Australia

Fourie, C D (1993) 'A New Approach to the Zulu Land Tenure System: An Historical Anthropological Explanation of the Development of an Informal Settlement', Unpublished PhD Thesis, Rhodes University, Grahamstown, South Africa

Fourie, C (1996) 'Land Tenure and Administration', Paper presented at *International Conference on Land Tenure and Administration,* Orlando, Florida

Fourie, C and van Gysen, H (1996) 'Land Management and Local Level Registries', *South African Journal of Surveying and Mapping,* Vol 23, No 6, pp353–359

Harrison, P (1992) 'The Policies and Politics of Informal Settlement in South Africa: A Historical Perspective', *Africa Insight,* Vol 2, No 1, pp14–22

Jackson, J (1996) 'Extending the South African Cadastral System Using a Mid-point Method', *South African Journal of Surveying and Mapping,* Vol 23, No 5, pp277–284

Landman, K (1996) 'Mangosuthu Technikon's Role in Developing a Curriculum for Land Administration in South Africa', Paper presented at *International Conference on Land Tenure and Administration,* Orlando, Florida

Municipal Ordinance of the Cape of good Hope, Juta (as amended) 20, 1974

Nichols, S E (1993) 'Land Registration: Managing Information for Land Administration', PhD Dissertation, Department of Surveying Engineering Technical Report No 168, University of New Brunswick, Fredericton, New Brunswick, Canada

Okpala, D (1992) 'Land Management and Aid Programs', *South African Journal of Surveying and Mapping,* Vol 21, No 5, pp262–272

O'Riordan, T (1971) *Perspectives on Resource Management,* Psion Ltd, London

Parliament of the Republic of South Africa (1994) *White Paper on Reconstruction and Development*, Government Gazette 353, 16085, November

Parliament of the Republic of South Africa (1996) *Constitution of the Republic of South Africa* (as adopted by the Constitutional Assembly on 8 May 1996), Constitutional Assembly, South Africa

Payne, G (1989) *Informal Housing and Land Subdivisions in Third World Cities: A Review of the Literature,* Centre for Development and Environmental Planning, Oxford Polytechnic, Headington, Oxford

Peil, M and Sada, P (1984) *African Urban Society,* John Wiley & Sons, Chichester, UK

Rene, O H (1996) 'Some Reflections Following the Conference Participants', E-Conference: *Addressing Natural Resource Conflicts through Community Forestry,* January–April 1996, FAO, 8 March

Republic of South Africa (1996) *Our Land. Green Paper on South African Land Policy*, Department of Land Affairs

UNCHS (1991) *Institutional/Legal Options for Administration of Land Development,* Urban Management Programme, Background paper, UNCHS, Nairobi, January 1991

UNCHS (1996) 'New Delhi Declaration: Global Platform on Access to Land and Security of Tenure as a Condition for Sustainable Shelter and Urban Development', *South African Journal of Surveying and Mapping,* Vol 23, No 5, pp291–304

Urban Foundation (1991) 'Informal Housing: The Current Situation', *Policies of a New Urban Future*, The Urban Foundation, Johannesburg

Williams, C W (1997) 'Partnerships, Power and Participation', *Habitat Debate*, Vol 3, No 1, pp15–16

Williamson, I (1991) 'Land Information Management at the World Bank', *The Australian Surveyor,* March, Vol 36, No 1, pp41–51

Part 4 Conclusions

Chapter 15

Evaluating the Experience of Brazilian, South African and Indian Urban Tenure Programmes

Donald A Krueckeberg and Kurt G Paulsen

Drawing generalized conclusions from three unique countries to imply universally applicable policy solutions ignores the diverse histories, institutions and politics of these countries. Policy formulation and evaluation should never be an abstract academic exercise, but rather rooted in the various interests, powers, political and social movements and institutions within a country. Earlier chapters of this book have traced the evolution of urban tenure programmes, with insights drawn from researchers and activists in each country trying to understand their experience. In this chapter we highlight from that analysis six key issues for further discussion (all quotations are from the conference papers on which the chapters are based and not from the chapters themselves).

Urban households and tenure security

For urban households, tenure security is not only constituted by title but is intricately linked with access to jobs, education, credit, health care, water and sanitation, and family and social networks.

Jobs drive the formation of squatter settlements nearly everywhere. The most effective programme of land redistribution in South Africa is the result of two Acts that jointly address rural housing and jobs: the Extension of Security of Tenure Act (ESTA) and the Land Reform (Labour Tenants) Act. These Acts secure the rights of tenants not only to land but also to services. In South Africa's rural traditions,

> *Services such as access to water, electricity and even schooling, are seldom public goods, but rather part of a farm worker's employment package. On the one hand*

> *this illustrates the tremendous social power that landowners still enjoy over their workers. On the other, the private provision of such services is a social resource that the South African government dare not destroy* (Roux, 1999, p10).

At least with the provision of water, these Acts make the deprivation of these services tantamount to eviction.

In São Paulo, the Cingapura programme is a scattered site programme of 90,000 families in 243 locations, begun in 1993. It is a programme of 'verticalization', that is, high density, multi-storeyed housing that is meant to avoid displacement and to preserve social and economic networks and sustain community cohesion (Engelbrecht et al, 1999). In Brazil

> *...all upgrading projects include funding for the installation of social infrastructure, including clinics, day-care centres etc and the establishment of job creation initiatives, including training. No funding is provided for the construction of houses, except in cases where relocation housing is to be constructed* (Engelbrecht et al, 1999, p13).

The contrast with South Africa is clear; here programmes are narrowly defined as infrastructure and housing. In Cape Town, Cross observes that the prevailing model is based on delivering a finished house and is not financially viable. She states that,

> *access to tenure through this effectively gold plated vehicle is falling further behind the rate of increase in demand, so that the informal process of access to land is accelerating as it outruns the tenure formalization process that is chasing it* (Cross, 1999, p15).

At Blaauwberg, on the northern periphery of Cape Town, households who had just received services and title moved out within two months because moving into one of these upgraded houses meant paying for services and social pressure to adopt an expensive urban lifestyle. Secure tenure thus rests on financial and social support systems. Without these, titles may actually prove to be less conducive to secure tenure than present informal arrangements and customary community property regimes.

SETTLEMENT UPGRADING THROUGH TENURE SECURITY

Settlement upgrading through tenure security should be viewed not only as improving the internal environment of informal settlements, but also as integrating settlements into the formal and legal housing and land economy. Researchers, policy-makers and activists need to think critically about the positive and negative consequences of increased formalization and marketization of the urban tenure process.

First, there is a need to recognize that informality does not mean a lack of social or economic organization of squatter settlements. They already operate fully as market processes. Almost all land invasions involve a planning process. In Brazil it is called the 'chain of interest' or 'invasion industry', a chain that usually starts with someone politically well-connected. Trees are cut and burned, land plots are marked, streets are planned, sometimes schools and health facilities too, families quickly build foundations, and several dozen shanties are built and occupied. As the risk of expulsion passes, concrete houses, shops and churches are built (Serre, 1999). Sometimes the developers will give away the first plots and later charge high prices for the established development. Government is then forced to provide some services (Fernandez, 1997). Soon, the original poorer settlers may feel overwhelming pressure to sell, as they are being priced out of their homes, a process known as 'market eviction'.

Makhatini describes a similar process of invasion and settlement for Cato Manor in Durban, South Africa (Huchzermeyer, 1999b, p14). First comes camouflaged or hidden squatting, clandestine building activity, much of which occurs at night. Eventually the shacks are numbered, lending a degree of legitimacy to the settlement. Open squatting takes place in the second stage, in which authorities declare an official moratorium. Potential newcomers are diverted to other camouflaged developments and squatter leadership becomes sophisticated in negotiating for official services. A third stage is marked by the mobilization of political parties in the settlement.

The process of being accommodated in an informal settlement in both Durban and Cape Town follows more or less the rural model. Outsiders approach a relative or contact. Only couples with children are normally acceptable. A plot is identified and the applicant is taken to the headman or chief who reviews the family history, behaviour and what payment is expected. The applicants then stay with the sponsors so that the neighbours can observe their behaviour while the new house is being built. A neighbourhood party celebrates completion and marks acceptance. Up to seven levels of permission may be required, with a written record. 'Maintaining tenure rights then depends on good moral behaviour' (Cross, 1999, p12). Tenure security for those born outside the community is essentially contractual, continuing to be conditional upon acceptable behaviour and meeting communal obligations. The question raised by these studies of the settlement process is the impact of tenure intervention. Tenure is already granted through a variety of community means. It is an open question whether formal title based tenure will necessarily bring more security. Another line of criticism points out the fluid nature of the informal settlement process over time. This fluidity is in many ways important to the residents and to freeze its structure prematurely by fixing boundaries of ownership may stifle rather than expedite development.

Second, formal planning standards have the effect of excluding poor populations. If the aim is to integrate squatter settlements into the formal economy, then either formal planning standards need to be relaxed, or tremendous financial assistance needs to be offered to poor residents. Formal planning and policy responses to squatter settlements often exist in a policy and legal framework that denies their existence. In 1975, the Indian state of Maharashtra passed the Maharashtra Vacant Lands (Prohibition of Unauthorized Occupation and Summary Eviction) Act, which made it illegal to occupy or build on any land without permission of the municipal commissioner. The prevailing policy was slum clearance and eviction. While the Supreme Court of India abolished this Act in 1985, Pimple and John comment on its policy towards slums,

> *State organized attempts towards tackling the issues of housing for the urban poor seem to be primarily rooted in a planner's dream of a 'clean and slum free city' rather than any real concern for the rights of the poor to quality housing facilities and basic amenities... Thus interventions are rejection oriented attempts to wish away the problem; instead of being rooted in an acceptance of the phenomenon as a legitimate result of the developmental processes and patterns of the city and subsequent empathetic and consistent action towards need-identification and the comprehensive enhancement of quality of life* (1999, p10–11).

In Delhi, 30 per cent of the 12.8 million residents live in squatter settlements and 62 per cent live in various informal housing subsystems, most on public land. Delhi first had a master plan in 1957, but the master plan made no provision for squatter settlements (Risbud, 1999). There is evidence that the high planning standards of the master plan effectively exclude large populations from housing supply. Similarly in South Africa, tenure reform and upgrading has not been viewed as a part of planning (Xaba and Beukman, 1999, p9). In Brazil elitist standards of development, like exclusionary zoning in the United States, set the bar of market entry far above affordability. The high value areas that result are driven higher by appreciation of real estate from public investments (Saule, 1999a).

Third, delivery of formal sector services triggers questions of taxation and payment. Ideally, well-designed property taxes and service charges can provide adequate municipal revenue for economic development, infrastructure and replicability of services. In practice, however, most squatter residents are either used to acquiring services and land without taxes and fees, or are unable to pay. Creative, community-based solutions to taxes and service fees will need to be found.

Non-payment can also be politically motivated. Current housing policy in South Africa assumes that ownership will foster civic pride. The provision of municipal services is seen as conditional on using that ownership to identify end users and thus ensure their payment for services. Yet, in practice, services payment levels in towns in the Greater Nelspruit area are at about 25 per cent. One explanation often given for this non-payment of services is that these (black) South Africans, in the latter days of apartheid, often withheld payment for services as a form of political protest. This pattern of non-payment persists after apartheid because people assume that the services are deserved as compensation for past deprivations and injustices. Another explanation is given by Cross:

> *It is still an open question as to how far non-payment of service charges in the impoverished informal areas is due to unemployment and lack of money, or to a stubborn rurally derived belief that services are part of the right to settle and represent a trade-off against basic needs once filled by the rural natural resources base, but now out of reach due to overcrowding and environmental damage caused by apartheid spatial planning* (1999, p7).

Fourth, integration into formal processes overrides and destroys customary and cultural systems of land tenure management. This is seen especially in tribal areas under urban encroachment where conflict arises. Traditional leaders and members press for legal recognition of customary and common tenure arrangements and fear that incorporation into formal systems will reduce tenure security. Traditional leaders also view tenure regularization as a threat to their political status. The process needs to be managed with sensitivity to cultural differences and with greater community participation in the process.

Tenure rules

In all three countries, tenure changes take place in a changing context of legal rules and enforcement. These rules range from international accords, national constitutions and provincial legislation, to local ordinances and policies reflecting the contest between historical property owners and those excluded from property rights.

Before the 1988 constitution of Brazil, the legal recognition of private property rights was nearly absolute, overpowering any notion of state control or 'social interest' (Fernandez, 1997). The 1988 constitution conferred great power on local authorities to guarantee the 'full development of the city's social functions'. It states that 'urban property only accomplishes its social function when it attends to the "fundamental requirements of city orderliness expressed in the master plan"' (Fernandez, 1997, p22). The constitution also ensured the possibility of popular participation in the local planning process. A non-governmental organization (NGO) now, for example, is able to submit a bill directly to the local legislature.

> *In the terms of the 1988 constitution the economic content of urban property is to be decided by municipal government through a participatory legislative process, and no longer by the exclusive individual interest of the owner* (Fernandez, 1997, p24).

He calls this the new 'right to urban planning'.

In addition the constitution recognized Brazil's commitment to a 'right to housing', as expressed in its commitment to numerous international accords (Saule, 1999a, p2). While these obligations are not interpreted to mean that the state must offer a residence to each citizen, they are understood to mean that the state cannot regress in housing law and is obliged to intervene and regulate the private sector in housing matters. While guidelines for land policy are a federal responsibility, states and municipalities have obligations to promote home building and improvements. Execution is purely a municipal responsibility, requiring the development of a local housing policy through a master plan.

In South Africa the fall of apartheid was followed by an interim constitution in 1993. Reform was constrained by a property clause similar to the fifth amendment of the constitution of the United States. It guaranteed the right to acquire and hold property, to dispose of those rights and not to be deprived of them except by due process for a public purpose with just compensation. The constitution was revised in 1996 to contain an elaborate and radically new property clause, including provisions for equitable land reform, secure tenure or compensation and restitution for land dispossession (Roux, 1999).

Under the Indian constitution (article 246), land policy, land supply, urban development and housing policies are primarily state concerns, not the concerns of the central government (Banerjee, 1999a). Thus, great variation can be observed in the land policies and responsiveness of different state governments to urban pressures. It should be noted, however, that many of the financing mechanisms available for urban development are federal responsibilities and all state governments are heavily dependent on the central government for administrative and financial resources. Most municipalities lack qualified administrative staff and resources, as most administrators of various policies are central government staff posted to states and cities. It becomes apparent that even though the central government has promoted the concepts of devolution and decentralization, much administrative and fiscal capacity remains at the national level.

The degree of civil society involvement varies between states, but there is a national campaign for housing rights. Additionally, under Indian law any party can file public interest litigation, hence there have been numerous court rulings that slum clearance is legitimate only under strict public interest concerns (for example health or infrastructure) and only with compensation or adequate relocation. These court actions have often served to improve de facto security of tenure.

Local institutions are a key variable in the success of upgrading programmes

In general, the number of institutions, both municipal government structures and NGOs, and the interaction between them are crucial determinants of programme success. Belo Horizonte's PROFAVELA programme was based on a 1979 federal enabling law. Its main features were to establish legal parameters for regularization of land subdivisions in favelas and to establish criteria for the transfer of local government land to the occupiers, respecting the original characteristics of the favelas as far as possible and recognizing a basic right to housing. The programme created the urbanization company of Belo Horizonte (URBEL). The URBEL was not very effective until 1993 when a new government created the municipal system of housing, the municipal council of housing and the municipal fund of popular housing. The city also has a participatory budget process and a municipal conference on housing every two years with 1000 delegates, to evaluate and revise housing policy (Bede, 1999).

In Recife, under an authority named COMUL, teams of technical staff worked with residents and lawyers to prepare an 'urban upgrading plan and the plan for land regularization' (Pinho, 1999, p6). The work involved NGOs whose professionals mediated between the public authority and the community organizations. In Brazilia, housing policy was begun through the organization of a district housing conference in 1996 in which some 10,000 people chose 703 delegates. The conference is held every two years. A housing council, which is political and sets policy, and a housing fund were established in parallel. In 1997 the government instituted a social interest housing programme (PHIS) to regularize low income housing in the district.

Intergovernmental coordination is very important. Brazil tends to work with strong political leadership driving projects and many agencies working closely together. South African development agencies tend to work in isolation and to deliver fragmented services. Indian agencies tend to be administrative arms of federal and state ministries. Titling and the registration of ownership in Brazil are not centralized nationally. 'The Deeds Registration system is accordingly highly decentralized and is separated from local government cadastral systems, planning frameworks and valuation rolls for the determination of rates and taxes' (Engelbrecht et al, 1999, p13). This inhibits the ability of local government to collect taxes and means that ownership is generally less important. South Africa has a highly sophisticated cadastral system, deeds office infrastructure and local government rating system, but this is complex, time-consuming and expensive and also fails to assure tax collection. In India, state revenue ministries are responsible for registering and titling plots.

South Africa offers a vivid example of the impacts of poor intergovernmental coordination in the programme in the Greater Nelspruit area. An assumption has been that ownership would be a catalyst for local economic development and would halt the practice of borrowers taking out life insurance policies as security for loans. Owners with titles, it was hypothesized, could now use their property as security, freeing the insurance payment monies for other investments. Unfortunately, at the same time that government has been upgrading tenure, it has been reducing investment in water and transport, decreasing public transport subsidies, raising commuting costs and hence causing consumers to divert savings from investment to water and transport. Reduced investment in water and transport are perceived to be a de facto reduction in tenure security. In addition there are new transfer or conveyancing costs incurred with ownership (Ambert, 1999). This demonstrates the lack of intergovernmental coordination of policies.

In contrast to both Mumbai and Delhi, Visakhapatman (Vizag) has seen remarkable achievements in slum improvements and regularization. Vizag shows that:

> *it is actually possible for diverse institutions with different activities to work in an integrated way, but this does not happen automatically. The process has to be backed by institutional development and clear procedures. The support of the State Government has been a crucial element in the process... The link between tenure regularization, housing and infrastructure is one of the success stories of slum improvement in Vizag* (Banerjee, 1999b, pp16–17).

Vizag has two decades of experience in slum improvement and tenure regularization, leading to improvements in slum conditions and access to infrastructure, health care and literacy. In 1979, the urban community development (UCD) programme started, focusing on community participation via neighbourhood committees to strengthen the capacity of slum dwellers. This was accomplished with a low programme budget. The UK Department for International Development (DFID) funded a city-wide slum improvement programme based on the principles of the UCD, with grants for infrastructure improvement and increased spending on social, economic and health activities. Community organizers played a key role in the decision-making, and UCD staff (of the municipal corporation) conducted surveys with grass-roots workers.

Gender is a central element of the tenure process, as urban tenure security programmes often exclude women

Urban researchers in the developing world are increasingly paying attention to gender and development issues; some argue that women's access to 'house and land' is a key determinant of women's empowerment in urban areas (Tinker and Summerfield, 1999). Mearns (1999) describes how many women are excluded from holding title to land through either legal or cultural means, and that this lack of access is a key determinant in women's economic status and poverty. In both India and South Africa, although women have some legal rights to own property, many cultural traditions deny women inheritance or management rights. The struggle for women's property rights requires legal, institutional and cultural transformation. In many cases, tenure reform without explicit concern for women can leave women more disadvantaged. Cross (1999) points out that upgrading is seen as competing with the private rental sector, also known as backyard shack farming (shacks for rent in the backyard to supplement the owner's income). Single women and women with children are often not welcome and have to make special rental arrangements as lodgers. The criticism appears to be that the introduction of titled tenure sharpens the status divisions between owners and renters and landlords and tenants, ironically reproducing the very dependency relationship that tenure security was designed to erase. Those who suffer most are women. In India, however, as a result of the active women's movements in civil society, specific legislation addresses these concerns. Innovations in the National Housing Policy Act of 1994 confer joint or exclusive title to poor and disadvantaged women (Mearns, 1999).

Actually, there is some evidence that programmes with strong women's leadership deliver more secure tenure. In Vizag there has been substantial gentrification and 'downward raiding' (the selling of plots and houses to higher income families). This is especially so in the well-located hills along the national highway. Rates of property value increases in the hills have surpassed even those in the central city. Many sales are considered distress sales because of the debt of the homeowner. The initial enthusiasm of building a nice home on secure land with secure tenure allowed too much debt at high interest rates, presumably from a variety of non-formal lenders. However, in well-organized neighbourhoods with women dominating local committees, the turnover of plots is much lower.

Corruption and Violence Are a Key Concern

In some international documents, there seems to be the assumption that simply the adoption of correct policies will improve the lives and position of slum dwellers. In reality, the whole process is often infused with violence, corruption and ethnic or class politics.

Social unrest can ensue when groups perceive unfairness or when middle income residents perceive unequal benefits to the poor. Poor residents can become violent when the process appears too slow or too arbitrary. Slum organizers, political bosses and tribal chiefs can often view tenure regularization as eroding their privileged social and economic position. Municipal officials and ministries that exhibited near absolute power over land decisions do not easily give up control. Political sympathy for squatters is frequently low. Change, which improves the situation of some, will necessarily erode political, cultural and/or economic power for others. For all these reasons and more, the process is often complicated, political and violent. The key challenge for policy-makers, planners and community organizers is to have a process that is as open, transparent and participatory as possible. Even then, we should not assume that the process will be easy.

In Brazil, the residents of favelas are, in the words of Castro (1999a), 'the hostages of a criminal state within a state'. In South Africa, the fall of apartheid and its policies of corruption and violence were believed to usher in a new era of stability and openness. Instead, the experience of Durban and Cape Town suggests that private market approaches have led to serious problems. Private developers get discouraged and walk away from unfinished projects; local organizations then collect money and walk away with it. In worse cases, as happened in Durban, these organizations resist completion of the projects, which would eliminate their roles, and sometimes even kill their critics. As Cross puts it, 'Emphasis on participatory process may be contributing to encourage local leaders to take advantage of their opaque and unaccountable positions to exploit their constituencies...' (1999, p17).

Conclusion

The underlying causal processes that generate the spatial pattern of slums and squatter settlements are many and complex. No simple, universal policy solution can solve all attendant problems. Programmes to improve tenure security are, therefore, an essential policy intervention, but must involve community participation and must be integrated into broader programmes of planning, policy, development and justice.

References and Bibliography

Alfonsin, B de Moraes (1999) 'Instruments, Experiences and Trends in Tenure Regularization in Brazil', Paper presented at the workshop on *Tenure Security Policies in South African, Brazilian, Indian and Sub-Saharan African Cities: A Comparative Analysis*, Johannesburg, 27–28 July

Allanic, B, Pienaar, J, Waring, M and Shilote, B (1999) 'Some Perception and Realities Observed During Recent Land Reform and Livelihoods Research of Eight Localities in the Eastern District of the North-West Province', Paper presented at the workshop on *Tenure Security Policies in South African, Brazilian, Indian and Sub-Saharan African Cities: A Comparative Analysis,* Johannesburg, 27–28 July

Ambert, C (1999) 'Privatising Displaced Urbanization: Can Tenure Reform Lead the Way for the Integration and Development of Displaced Urban Areas in a Neo-liberal Environment? Issues Emerging from the Greater Nelspruit Tenure Upgrading Process', Paper presented at the International Research Group on Law and Urban Space/Centre for Applied Legal Studies workshop on *Facing the Paradox: Redefining Property in the Age of Liberalization and Privatisation,* Johannesburg, 29–30 July

Banerjee, B (1999a) 'Security of Tenure in Indian Cities: Overview of Policy and Practice', Paper presented at the workshop on *Tenure Security Policies in South African, Brazilian, Indian and Sub-Saharan African Cities: A Comparative Analysis*, Johannesburg, 27–28 July

Banerjee, B (1999b) 'Security of Tenure in Irregular Settlements in Visakhapatman', Paper presented at the workshop on *Tenure Security Policies in South African, Brazilian, Indian and Sub-Saharan African Cities: A Comparative Analysis*, Johannesburg, 27–28 July

Bede, M M C (1999) 'Housing Policy for Low Income Population of Belo Horizonte', Paper presented at the workshop on *Tenure Security Policies in South African, Brazilian, Indian and Sub-Saharan African Cities: A Comparative Analysis*, Johannesburg, 27–28 July

Berrisford, S (1999) 'Redistribution of Land Rights in South Africa: Simply a Rural Question?', Paper presented at the International Research Group on Law and Urban Space/Centre for Applied Legal Studies workshop on *Facing the Paradox: Redefining Property in the Age of Liberalization and Privatisation,* Johannesburg, 29–30 July

Castro, S Rabello de (1999a) 'Favela Bairro: A Brief Institutional Analysis of the Programme and its Landed Aspects', Paper presented at the workshop on *Tenure Security Policies in South African, Brazilian, Indian and Sub-Saharan African Cities: A Comparative Analysis,* Johannesburg, 27–28 July

Castro, S Rabello de (1999b) 'Some Different Ways of Thinking and Reconstructing Property Rights, and Land Tenure Rights in "New Countries"', Paper presented at the International Research Group on Law and Urban Space/Centre for Applied Legal Studies workshop on *Facing the Paradox: Redefining Property in the Age of Liberalization and Privatisation,* Johannesburg, 29–30 July

Cross, C (1999) 'Land and Security for the Urban Poor: South African Tenure Policy under Pressure', Paper presented at the workshop on *Tenure Security Policies in South African, Brazilian, Indian and Sub-Saharan African Cities: A Comparative Analysis,* Johannesburg, 27–28 July

Engelbrecht, C, Chawane, H, Gollocher, R, van Rensburg, G, Mkhalali, X, van der Heever, P and Browne, C (1999) *Upgrading Informal Settlements: A Brazilian Experience,* Gauteng Department of Housing, Johannesburg, 1 July

Fernandez, E (1997) *Access to Urban Land and Housing in Brazil: 'Three Degrees of Illegality',* Lincoln Institute of Land Policy Working Paper, Cambridge

Huchzermeyer, M (1999a) 'A Case Study of the Weilers Farm Informal Settlement Process: Questions Raised about the Current Intervention Framework', Paper presented at the International Research Group on Law and Urban Space/Centre for Applied Legal Studies workshop on *Facing the Paradox: Redefining Property in the Age of Liberalization and Privatisation,* Johannesburg, 29–30 July

Huchzermeyer, M (1999b) 'Tenure Questions as Understood in the Debates in South African Informal Settlement and Intervention Literature of the 1990s: A Critical Review', Paper presented at the workshop on *Tenure Security Policies in South African, Brazilian, Indian and Sub-Saharan African Cities: A Comparative Analysis,* Johannesburg, 27–28 July

Imparato, E (1999) 'Security of Tenure in Brazil', Paper presented at the workshop on *Tenure Security Policies in South African, Brazilian, Indian and Sub-Saharan African Cities: A Comparative Analysis,* Johannesburg, 27–28 July

Kuhn, J (1999) 'Customary Tenure in South Africa: The Final Frontier', Paper presented at the workshop on *Tenure Security Policies in South African, Brazilian, Indian and Sub-Saharan African Cities: A Comparative Analysis,* Johannesburg, 27–28 July

Lyngdoh, Y R (1999) 'Effect of Tribal Land Tenure Systems on Urban Development: Case Study of Shillong, North-East of India', Paper presented at the workshop on *Tenure Security Policies in South African, Brazilian, Indian and Sub-Saharan African Cities: A Comparative Analysis,* Johannesburg, 27–28 July

Mearns, R (1999) *Access to Land in Rural India: Policy Issues and Options,* World Bank discussion paper, available at www.worldbank.org/research

Pimple, M and John, L (1999) 'Security of Tenure: the Experience of Mumbai', Paper presented at the workshop on *Tenure Security Policies in South African, Brazilian, Indian and Sub-Saharan African Cities: A Comparative Analysis,* Johannesburg, 27–28 July

Pinho, E (1999) 'The Legislation of Social Interest in Brazil', Paper presented at the workshop on *Tenure Security Policies in South African, Brazilian, Indian and Sub-Saharan African Cities: A Comparative Analysis,* Johannesburg, 27–28 July

Risbud, N (1999) 'Policies for Tenure Security in Delhi', Paper presented at the workshop on *Tenure Security Policies in South African, Brazilian, Indian and Sub-Saharan African Cities: A Comparative Analysis,* Johannesburg, 27–28 July

Roux, T (1999) 'Something Less than Ownership: Extending Security of Tenure in Rural South Africa', Paper presented at the International Research Group on Law and Urban Space/Centre for Applied Legal Studies workshop on *Facing the Paradox: Redefining Property in the Age of Liberalization and Privatisation,* Johannesburg, 29–30 July

Saule, N, Jr (1999a) 'Forms of Protection of the Right to Housing and the Combat of Forced Eviction in Brazil', Paper presented at the International Research Group on Law and Urban Space/Centre for Applied Legal Studies workshop on *Facing the Paradox: Redefining Property in the Age of Liberalization and Privatisation,* Johannesburg, 29–30 July

Saule, N, Jr (1999b) 'Urban Land Regularization: The Experience of Brazil', Paper presented at the workshop on *Tenure Security Policies in South African, Brazilian, Indian and Sub-Saharan African Cities: A Comparative Analysis,* Johannesburg, 27–28 July

Serre, A (1999) 'Illegal City and Urban Policy in Belém, Brazil', Paper presented at the International Research Group on Law and Urban Space/Centre for Applied Legal Studies workshop on *Facing the Paradox: Redefining Property in the Age of Liberalization and Privatisation,* Johannesburg, 29–30 July

Tinker, I and Summerfield, G (1999) *Women's Rights to House and Land: China, Laos, Vietnam,* Lynne Rienner Publishers, Boulder, Colorado

Xaba, S and Beukman, R (1999) 'The Upgrading of Informal Tenure in South Africa: the Policy Gaps and the Legislative Challenges', Paper presented at the workshop on *Tenure Security Policies in South African, Brazilian, Indian and Sub-Saharan African Cities: A Comparative Analysis,* Johannesburg, 27–28 July

Chapter 16

The Experience of Tenure Security in Brazil, South Africa and India: What Prospects for the Future?

Alain Durand-Lasserve and Lauren Royston

CURRENT TRENDS REGARDING SECURITY OF TENURE

Regularization projects and programmes in the 1990s have made significant progress in responding to the demand from the poor for land and housing, in a context of public and private formal sector failings (Urban Management Programme, 1995a; UNCHS, 1996a). Previous government attitudes and responses to the expansion of irregular settlements were inadequate, based mainly – depending on the cities – on a combination of repressive options and laissez-faire policies. Although the provision of serviced land and self-help practices have helped to curb the pressure of housing demands, they have by no means met the needs of low-income populations. Conventional tenure regularization policies, as implemented in many developing cities over the last decade, have proved to constitute a better response to the needs of the urban poor. These policies rely on two sets of interrelated measures: on the one hand, the de jure recognition of de facto occupancies, and on the other, the provision of services to the settlements concerned.

The chapters in this collection support the evidence that the needs of poor urban populations can best be met by the integration of irregular settlements, not because this is technically the best solution, but because it is simply the only one available (Durand-Lasserve and Clerc, 1996). The accelerated commodification of all the popular informal land and housing delivery systems and the disengagement of the state as producer of land and housing are the result of market forces (Baross and van der Linden, 1990). These are legitimized and enforced by the prevailing neo-liberal economic orthodoxy.

However, tenure regularization policies based on the allocation of individual freehold titles clearly have limitations (UNCHS, 1993; Payne, 1997; Fourie, 1999). The reasons for this are many and varied, and are demonstrated in several chapters in this book. A similar set of obstacles in the implementation of such measures is discernible, however different these may be, in both national and local contexts:

- the contradiction between the right to housing and the right of ownership;
- lack of human and financial resources;
- the poverty of the populations concerned when any attempt is made to set up cost recovery measures;
- ill-adapted regulatory frameworks and land management procedures in a context of legal complexity;
- lack of appropriate land-related information systems and land records;
- over-centralized land management and registration procedures; and
- little community participation.

It has become clear, at a global level, that access to security of tenure for the urban poor through formal registration and the mass provision of individual property titles is rarely possible (for technical, political, administrative, financial and cultural reasons), nor is it always desirable. It was argued in the introduction that the actors for whom the provision of freehold titles is essential are formal private developers. For populations living in irregular settlements, as well as for governments, more flexible options combining protection measures and leases are a possibility, especially on publicly owned land (Payne, 1999). However, the demand for property titles still predominates, mainly for historical and cultural reasons, or because administrative traditions and bureaucratic vested interests hinder the adoption of alternative tenure options. As stressed by Fourie,

> *Rights such as freehold and registered leasehold, and the conventional cadastral and land registration systems (paper or digital), and the way they are presently structured, cannot supply security of tenure to the vast majority of the low income groups and/or deal quickly enough with the scale of urban problems, and innovative approaches need to be developed* (1999).

Regularization policies implemented to date have, with a few rare exceptions (Mexico in the 1990s), achieved only limited results (Azuela, 1998; Varley, 1999). They waver between forms of blanket mass regularization, which have particularly benefited middle income groups living in unauthorized land developments, and pilot projects, which have reached only a small fraction of the populations concerned.

Security of tenure must be seen as a prerequisite, or an initial step, in an incremental tenure regularization process, focusing particularly as it does on the protection, as opposed to the eviction, of the irregular settlement occupants, and not on their immediate regularization in legal terms. Approaches that try to achieve security of tenure are the only ones that will meet the immediate and the longer-term needs of the populations. However, all the contributions in this book show that governments are still unwilling to commit themselves to this option. Paradoxically, it requires very little, and yet too much, effort on their part. Little, in that, in the initial stages at least, it does not require a large amount of financial input, nor the establishment of long, complicated procedures like those involved in the formal registration and the allocation of property titles. Existing community organizations may even be willing to assist. Too much, in that it presupposes a clearly declared political will (and the means to enforce it) and a form of government that is willing to listen to popular claims and recognize their legitimacy, while accepting the political risks and cultural changes involved (Leckie, 1992).

Resistance on the part of many urban stakeholders to security of tenure policies is best explained either by economic concerns or by the category of actors to which they belong (eg landowners whose interests have been harmed, civil servants who have lost their administrative power and influence, etc). However, most resistance is from cultural and ideological factors. Initial demands by the populations are usually for individual freehold title. The rapid allocation of real rights seems – often with good reason – the sole and ultimate protection against the combined pressures of the market and the administration, although experience demonstrates that this can only be a long-term objective. However, indications of change are evident. In India there is a clear revival of interest in ways to simply protect the rights of occupants with a gradual progression to the allocation of property rights. This kind of interest is emerging in Brazil and to a lesser extent in South Africa, where the historical denial of ownership to black people has influenced its standing as the preferred form of tenure.

Governments increasingly consider evictions as an inappropriate option (UNCHS, 1996b; Cobbett, 1999, pp1–6). They do not resolve the problem of access to land and housing for the urban poor. The political cost of evictions is often high, both internally and at international level. International debate and the sharing of experiences in the preparation of the 1996 Habitat II Conference contributed considerably to this development.

Change in government attitudes regarding access to property through adverse possession has been a long, slow process. Different countries are progressing at different rates, but progress is none the less evident. The position of the state in the process of land allocation and the creation of property is at stake. Private property is created either 'from the top', through land allocation by the state, or 'from the bottom', through adverse possession (Comby, 1995). For years, governments have reluctantly admitted that adverse possession procedures could benefit poor urban communities. In the absence of any alternative, procedures based on adverse possession are now increasingly considered as a realistic option.

The relevance of the dominant model is being questioned in the approaches by aid and cooperation agencies and to a lesser extent by international finance institutions (World Bank, 1999). International institutions involved in urban development cooperation programmes are increasingly aware of the perverse social effects of their aid and lending policies, and especially of the limitations of urban development strategies based predominantly – if not exclusively – on market mechanisms. They are now attempting to reassess their strategies and redefine priorities in a more pragmatic way, taking into account the diversity of local situations and needs. The same debate is occurring in several bilateral aid agencies and institutions responsible for cooperation with developing countries in the urban sector (see, for example, ISTED, 1998). Subsidy issues, the need for local land management and the recognition of the diversity and legitimacy of tenure systems are at the forefront of these debates.

The right to housing is increasingly being recognized at international level and all states are now beginning to attach importance to this right. New national constitutions and recent constitutional amendments in many countries are a reflection of this trend. In 1992 Leckie observed that:

> *The vast majority of national constitutions neither include the right to housing nor mention policies or measures towards improving housing conditions. Because national constitutions are generally the highest law of the land, the importance of including the right to housing as well as de jure and de facto protection against forced eviction cannot be overemphasized.*

The situation has evolved considerably over the past few years, often in parallel with the democratization process. A growing number of states now recognize, in principle at least, the importance of the notion of the right to housing, and this is true in a wide range of regional and cultural contexts. Implementation and enforcement remain the main problem, especially where land, housing and planning responsibilities have been transferred to state, provincial or municipal authorities.

> *The active enforcement of human rights in general, and housing rights in particular, is not an easy matter. In some cases it is relatively simple but, in most instances, there are difficulties. Many reasons contribute to this, not least of which is that the notion of state sovereignty is still the driving force behind national law* (Leckie, 1992).

State sovereignty was at the forefront of debates on the right to housing at the Habitat II Conference.

Various forces have converged at the international level to champion the cause of the right to housing. The Habitat Agenda (UNCHS, 1996b) and the Istanbul Declaration affirm '... the right to adequate housing as set forth in the Universal Declaration of Human Rights...' (ibid, section 26) and '...commitment to the full and progressive realization of the right to adequate housing, as provided in international instruments...' (ibid, section 39). Recognition is given to '...an obligation by governments to enable people to obtain shelter and to protect and improve dwellings and neighbourhoods'. One of the objectives (ibid, section 40) is to 'provide legal security of tenure and equal access to land to all people, including women and those living in poverty'. Another is to ensure '... transparent, comprehensive and accessible systems in transferring land rights and legal security of tenure'. However, the Habitat Agenda is not totally free of contradictions. On the one hand, there is an indication of the difficulties involved in implementing a policy that aims to promote the mechanisms of the marketplace (ibid, section 71, 'Enabling markets to work'). On the other, there is recognition of the need to make necessary adjustments: 'Government interventions are required to address the needs of disadvantaged and vulnerable groups that are insufficiently served by markets.'

Habitat International Coalition and various regional initiatives (such as the Asian Coalition for Housing Rights) have played a major role in the global movement for housing rights since the early 1990s. This role has since been taken on by United Nations organizations, the United Nations Centre for Human Settlements (UNCHS) in particular, and by inter-institutional programmes such as the urban management programme and the municipal development programme.

However, realization and fulfilment of human rights principles remains a major problem. The fact that the right to housing is implicitly or explicitly recognized in state constitutions does not necessarily mean that it is implemented and enforced. The right to housing may be in conflict with other principles (eg the right to property), economic constraints and administrative or court practices. This question is taken up in most chapters of the book, especially in the chapters on Brazil and South Africa.

In Brazil, as Saule stresses in Chapter 8,

> *with some ten years having passed since the promulgation of the Brazilian constitution, the following questions arise: how to guarantee compliance with the social function of property in the solution of urban environmental conflicts, considering the increase in the number of forced evictions in urban areas … by federal, state and municipal government, and by the judiciary and the institutions essential to the administration of justice.*

In the case of South Africa, Berrisford (1999) discusses this issue from another perspective:

> *At the heart of any legislative framework for planning and development lies the question of protecting individual property rights. To what extent can the state diminish rights – or interests – enjoyed by landowners in its endeavours to meet broader development and environmental needs?*

The case studies stress the difficulty, if not the impossibility, of operating with a centralized system of land management and registration. The recognition of land rights, the identification of beneficiaries, the supervision and management of the change in legal status and service provision can only be undertaken locally. However, the capacity of local authorities is still hindered by a lack of resources, both human and financial. In India, Brazil and South Africa, despite considerable devolution of land management responsibilities, municipalities are heavily dependent on central government resources for implementing large-scale tenure upgrading programmes. Except in Brazil, registration procedures and the allocation of real rights remain the prerogative of central government administrations. The decentralization of responsibility for land management would seem to be a sine qua non for the implementation of tenure security policies: first, so that the problem of irregular settlements is handled by local authorities and, second, so that the populations concerned can be involved in the process. This requires a redefinition of the roles and responsibilities for strategic planning and financial and technical support, and a high degree of delegation of responsibilities to municipal level. However, without sufficient resources at the disposal of local authorities, local implementation of tenure security policies will remain seriously flawed.

The case studies also emphasize the key role of community participation in the tenure upgrading process. This reflects both the shift away from a centralized decision-making process to a decentralized one, and the emergence of civil society organizations in planning and development projects. A close link exists between the democratization process at national and local levels, and the recognition of low-income communities (Huchzermeyer, 1999). In addition there are strong technical reasons motivating support for community participation in the tenure upgrading process. Procedures such as the identification of rights holders and beneficiaries, the resolution of land-related conflicts, adjudication, land allocation, cost recovery and project follow-up are more effective with community participation, rather than exclusive reliance on municipal officials.

An important lesson from the case studies is that innovation lies not in new tools, but in the new uses to which existing tools can be put. As far as land management is concerned, technical innovations are limited mainly to land information systems and cadastral techniques (dramatic simplification in survey techniques) and to new forms of contractual tenure arrangements. Changes in the arenas of planning (shifts to strategic planning) and finance (setting up micro-finance procedures adapted to the needs and resources of the urban poor) may, to a lesser extent, create a favourable environment for the emergence of innovative land management. In all three countries the main feature of social change is the consolidation of the democratic process and the increasing pressure exerted by civil society on decision-making processes at settlement and, more importantly, at city level. Innovative policies regarding security of tenure are the result of a change in the balance of power between urban stakeholders.

Some of the most interesting innovative practices include:

- The implementation of a wide range of tenure options, providing security of tenure with or without real rights and including non-transferable leasehold (*Patta*). Indian states offer several examples of such leases.
- Land sharing, as implemented in Hyderabad, or land reservations for the poor in Madhya Pradesh and Tamil Nadu.
- Designated 'special zones' for tenure upgrading or regularization, such as the 'special zones of special social interest' (ZEIS) in Brazilian cities, formally incorporate irregular settlements into the city fabric and protect them against arbitrary forced removal (although this dates back to the early 1980s, its application in the context of the 1988 constitution offers

interesting perspectives). In Indian cities, the practice of declaring illegal subdivisions as slums under the Slum Act provides communities with security of tenure and means that subsidized services can be extended into their settlements under the 'environmental improvement of urban slums' (EIUS).
- The notion of the 'social function of urban property', the authority of the master plan, along with the formal recognition of participatory processes in the Brazilian constitution of 1988.
- Adverse possession as a tool for securing residential tenure. India and Brazil offer several examples.

MAIN CHALLENGES REGARDING SECURITY OF TENURE

In general terms, a contradiction exists between the two stated objectives of 'ensuring access to land for all', and 'enabling the housing market to work'. This contradiction, reflected in the Habitat Global Plan for Action (UNCHS, 1996b), is clearly evident in the three countries discussed here. It has arguably been reinforced recently with increased inequalities and further imbalances between regions. The shift in economic policy in India, as the country moves towards a more liberal stance in the Eighth and Ninth Five Year Plans (1992–2002), the financial, fiscal and social crisis in Brazil and the affirmation of neo-liberal macro-economic framework in South Africa have all contributed to this situation (Huchzermeyer, 1999). Interventions by public authorities rarely compensate for the adverse social effects that have resulted. Such intervention mainly takes the form of housing subsidies in South Africa and attempts in India to find new forms of public–private partnership (Mahadevia, 2000), with improved popular participation in the decision-making process at municipal level. However, the ability of public authorities to regulate and integrate a system of access to land and housing, whether at central or local level, is constantly hindered by a lack of available resources.

Contradictions are also found between constitutional precepts, central and local government policy and practice, community practices and court interpretations. This problem, which can often lead to bottlenecks in the decision-making process, affects all three countries. It is an over-simplification to reduce it to a question of coordination between levels of government. Devolution of new powers to the municipalities in respect of land and housing, unaccompanied by the transfer of appropriate human and financial resources, is a contributing factor. It is the weight of community organizations and associations representing civil society that can make the difference when adjudication is needed.

Land delivery systems that offer a reasonable degree of security are still too firmly based on sectoral approaches. Security of tenure policies cannot be restricted to land policies. Other dimensions of the problem have to be taken into account, including community mobilization, community development, access to credit, training and capacity building, economic development and poverty alleviation programmes, and the integration of tenure upgrading and regularization policies into a planning approach at city or regional level. India and, to a lesser extent, Brazil offer examples of such approaches. Attempts at more integrated approaches are, however, hindered by economic and administrative factors.

Improved security of tenure often impacts negatively on the rental sector in informal settlements. This is a major problem, evoked by several authors, for which no satisfactory answers have yet been provided. Housing produced informally or illegally is often used for rental, where non-compliance with standards combined with insecurity of the settlements make it possible to produce dwellings at a rent affordable to low-income households. Improving security of tenure in a given settlement inevitably raises two questions. The first relates to identifying the beneficiary households: is it the owner of the land or the occupant of the dwelling who should be regularized? The response is clearly political. The second question concerns the link between tenure upgrading and market eviction. Tenants are in the front line here. Administrative measures aimed at improving or controlling the low-income rental sector may have undesirable adverse effects, and generate severe housing shortages by restricting the rental option.

Women are particularly disadvantaged in gaining access to land and securing tenure. Women, and especially single women with children, make up a very heavily disadvantaged group in terms of access to housing and security of tenure in all the countries considered. This question is considered in South Africa (by Cross in Chapter 12) where their extreme vulnerability is emphasized, especially in the informal rental sector, and in India (Banerjee, Chapter 2). The role of women in community-based organizations (CBOs) and associations often enables them to limit the effects of exclusion. Participatory schemes can improve their access to land and housing, which legal or administrative measures alone are unable to do (in any case these are few in number and difficult to enforce).

In the three countries, the existing legal and regulatory frameworks do allow, in principle, for the provision of secure tenure, although this may be difficult to implement and enforce. Enforcement by government institutions in charge of land management and registration, planning agencies and courts deserve particular attention. Four main cumulative reasons exist for the enforcement problem:

1 opposition of interest groups;
2 lack of political will;
3 lack of coordination between administrations in charge; and
4 lack of adaptation of the judicial systems in cities to the context.

In the case of South Africa, the gap between implementation and policy is particularly evident in the absence of alternative forms of tenure, despite policy commitment to choice and the availability of alternative legal instruments such as initial ownership. Court decisions are frequently in contradiction with government strategies regarding security of tenure. As observed by Fernandes in relation to Brazil in Chapter 6,

> *although more and more urban and environmental laws have been enacted over the past decades, especially at municipal level, their effective enforcement has been seriously undermined by the perseverance of several political, financial, institutional and legal factors. The slowness of the judicial machinery cannot respond to needs.*

As noted by Banerjee in Chapter 2, '"stay orders" and litigation are known to hold up development programmes, including housing projects for the poor, for years, with judgements being challenged and cases moving to higher courts'.

Providing security for households living in informal settlements requires precise information on the legal status of land and residential tenure, and up-to-date records of sales and transfers. In all three countries, formal land records and registration exist only for regular settlements. The problem is not purely technical in nature. Any tenure upgrading project requires, beforehand, a detailed land survey, the identification of rights holders and information about the exact nature of their rights. Such information can be collected only with the assistance of the communities concerned; a complicated and time-consuming process, which combines land survey (a technical activity) with negotiation and conflict resolution. Once the information has been recorded, and/or at a later stage registered, it must be maintained and updated. Conventional land registration procedures are not adapted to the situation and needs of irregular settlements, especially when they are centralized, as is the case in India and South Africa, and when CBOs are not closely involved in the process.

Where to from here? Some strategic guidelines for implementing security of tenure policies

Anti-eviction laws are a priority. It is now increasingly acknowledged that this must be the earliest priority, as protection from eviction is a prerequisite for any further action regarding tenure upgrading and regularization. Although an increasing number of countries have passed legislation protecting households in irregular settlements from arbitrary eviction, very few have adopted anti-eviction laws enforceable at national level. In this matter, South African legislation is fairly innovative, although its enforcement raises a series of problems. Situations vary greatly according to whether the illegally occupied land is publicly or privately owned. One of the major problems to be overcome is the contradiction between anti-eviction laws and the protection of property rights. It must be noted that, in any case, anti-eviction laws have limited effect if not combined with other integration measures.

Tenure upgrading and security of tenure measures should be accompanied by the provision of basic urban services. The provision of services should not be linked to occupancy status. In India, the draft national slum policy embodies the principle that 'households in all urban informal settlements should have access to certain basic minimum services, irrespective of land tenure or occupancy status' (Government of India, 1999). This principle is recognized by very few central and local governments.

It is now widely accepted that a wide variety of alternative tenure options should be available to meet the diversity of household needs in informal settlements. These options must be assessed and evaluated, depending on the specificity of local context. Any measure that could hinder the development of a particular land delivery subsystem may increase the pressure on other subsystems. The role of the public authorities should be to regulate the relationship between land and housing delivery subsystems and actors. Security of tenure, ensuring basic improvements and protection against displacement without the provision of suitable alternatives, should be an immediate, though temporary, objective. It should be a renewable but temporary response to the needs of the poor, a step towards more permanent status (lease or ownership), especially in contexts where formal tenure regularization cannot be undertaken on a large enough scale. Provision of individual property titles should be considered as a long-term, not immediate, objective.

Alternatives to individual ownership should be promoted. Conventional tenure regularization programmes based on the allocation of individual freehold are neither possible nor desirable. As noted by Fourie (1999),

> *this has led to problems of middle class downgrading, lack of affordability (especially when legalization is accompanied by service provision) and slow centralized delivery approaches... Instead, other approaches need to be considered. Given the bundle of rights associated with land and the different types of leases and rights available in different countries, it is suggested that ... leases become the instrument of choice for publicly owned land, and especially local authority land, rather than freehold. Private landowners should be encouraged to set up lease contracts with occupiers.*

Consideration should also be given to intermediate options, such as the provision of collective titles at settlement level. Support should be given to educate beneficiaries of tenure security programmes about the advantages of alternatives to ownership.

As it is rarely possible to provide freehold titles to households living in irregular settlements, it is essential that initial tenure rights be provided, in order to protect residents against eviction. These rights should be incrementally upgraded in accordance with what residents and governments need and can afford. Innovative land surveying and land registration in Namibia (Christiensen et al, 1999) illustrates such an approach, as does 'initial ownership' in South Africa, although it has never been implemented. Among other advantages, incremental processes tend to limit the effects of formal market pressures on the upgraded settlement. Ensuring a continuum between initial rights and freehold is also an essential objective. Parallel property registration systems must remain compatible with one another. A registration system providing 'second rank' titles for the urban poor should be avoided if such a system cannot be integrated into the formal registration system at a later stage.

Centrally registered rights such as freehold cannot meet the demand for security of tenure, first, for technical reasons and, second, because security of tenure is guaranteed when accepted both by public authorities and by the community at settlement and neighbourhood levels. Whenever possible, land recording and registration systems should be carried out at local level, rather than being centralized at national government level. Fourie (1999) stresses the move away from individual titles and towards a form of group title, with individual rights administered by the group members themselves, sometimes in partnership with the local authorities.

Tenure security should be a component of integrated approaches to development. Tenure upgrading should be integrated into strategic planning practices of municipalities. Any tenure upgrading or regularization programme should be accompanied by the provision of infrastructure and basic urban services. If provided at reasonable cost, services can consolidate the tenure upgrading process.

New forms of partnership that include informal actors should be pursued. The success of security of tenure programmes is heavily dependent on the capacity of communities to streamline individual and sometimes conflicting demands. This requires the consolidation of community organizations and the implementation of innovative forms of partnerships, including those with informal actors (Payne, 1999). It also requires the capacity to provide legal assistance to the poor. As stressed by Leckie (1992), such assistance should emphasize collective demands and group interests, encourage participatory involvement of potential beneficiaries and develop methods of group advocacy.

References

Azuela, A (1998) 'Tenure Regularization, Private Property and Public Order in Mexico', in E Fernandes and A Varley (eds) *Illegal Cities: Law and Urban Change in Developing Countries,* Zed Books Ltd, London, pp157–171

Baross, P and van der Linden, J (eds) (1990) *The Transformation of Land Supply Systems in Third World Cities,* Aldershot, Avebury, UK

Berrisford, S (1999) 'Urban Legislation and the Production of Urban Space in South Africa', Paper presented at the international conference *South to South: Urban–Environmental Policies and Politics in Brazil and South Africa,* Institute of Commonwealth Studies, London, 25–26 March

Christiensen, S F, Hoejgaard, P D and Werner, W (1999) *Innovative Land Surveying and Land Registration in Namibia,* DPU–UCL Working paper 93, May

Comby, J (1995) 'Comment fabriquer la propriété?' *Etudes Foncières,* Vol 66, pp28–35

Durand-Lasserve, A assisted by Clerc, V (1996) *Regularization and Integration of Irregular Settlements: Lessons from Experience,* Urban Management Programme, Working Paper Series 6, UNDP/UNCHS/World Bank UMP, Nairobi

Fourie, C (1999) *Best Practices Analysis on Access to Land and Security of Tenure,* Research Study, United Nations Centre for Human Settlements (Habitat), forthcoming UNCHS publication

Government of India, Ministry of Urban Affairs and Employment (1999) *Draft National Slum Policy,* New Delhi, January

Cobbett, W (1999) 'Towards Security of Tenure for All', *Habitat Debate*, Vol 5, No 3

Huchzermeyer, M (1999) 'The Exploration of Appropriate Informal Settlement Intervention in South Africa: Contributions from a Comparison with Brazil', PhD Thesis, Department of Sociology, University of Cape Town

ISTED (Institut des Sciences et des Techniques de l'Equipement) (1998) *Dynamique de l'urbanisation de l'Afrique au sud du Sahara,* Sous la direction de Michel Arnaud, Groupe de Travail Mécanismes et Logiques de l'Urbanisation en Afrique au sud du Sahara. Ministère des Affaires Etrangères – Coopération et Francophonie

Leckie, S (1992) *From Housing Needs to Housing Rights: An Analysis of the Right to Adequate Housing under International Human Rights Law*, International Institute for Environment and Development (IIED), Human Settlements Programme, London

Leckie, S (1995) *When Push Comes to Shove: Forced Evictions and Human Rights*, Habitat International Coalition, p139

Mahadevia, D (2000) 'Historical Note on Urban Policies', in *Urban Agenda in the New Millennium, India,* Report of a workshop held by the School of Planning, Centre for Environmental Planning and Technology, Ahmedabad, July 23–24

Payne, G (1997) *Urban Land Tenure and Property Rights in Developing Countries: A Review,* Intermediate Technology Publications/Overseas Development Administration (ODA), London

Payne, G (ed) (1999) *Making Common Ground: Public–private Partnerships in Land for Housing,* Intermediate Technology Publications, London

UNCHS (United Nations Centre for Human Settlements) (Habitat) (1993) *Evaluacion de las politicas nacionales del suelo e instrumentos para mejorar el acceso y el uso de la tierra urbana en America Latina,* Serie Gestion del Suelo, UNCHS, Nairobi

UNCHS (1996a) *Access to Land and Security of Tenure as a Condition for Sustainable Shelter and Urban Development,* Preparation document for Habitat II Conference, New Delhi, India, 17–19 January

UNCHS (1996b) *The Habitat Agenda: Goals and Principles, Commitments and Global Plan for Action,* Habitat II, UNCHS, Nairobi

Urban Management Programme, Nairobi, Ministry of Foreign Affairs, France, GTZ, Germany (1995a) 'Lessons from the study and the Seminar', *Urban land management, regularization policies and local development in Africa and the Arab States,* Land Management Series 4, UNCHS, Abidjan, 21–24 March

Varley, A (1999) 'New Models of Urban Land Regularization in Mexico: Decentralization and Democracy vs. Clientilism', Paper presented at N-aerus international workshop on *Concepts and Paradigms of Urban Management in the Context of Developing Countries,* Venice, 11–12 March, http://obelix.polito.it/forum/n-aerus

World Bank (1999) *A Strategic View of Urban and Local Government Issues: Implications for the Bank,* Transportation, Water and Urban Development Department, Urban Development Division, Draft, World Bank, Washington, DC

Index

Note: Page numbers in *italics* refer to boxes or tables in the text